Humusphere

HUMUSPHERE

Humus: A Substance or a Living System?

Herwig Pommeresche

Translated by Paul Lehmann

ACRES U.S.A.
GREELEY, COLORADO

Humusphere

© 2014, 2017, 2019 Organisher Landbau Verlag Kurt Walter Lau
Translation to English © 2019 Acres U.S.A.

All rights reserved. No part of this book may be used or reproduced without written permission except in cases of brief quotations embodied in articles and books.

The information in this book is true and complete to the best of our knowledge. All recommendations are made without guarantee on the part of the author, and Acres U.S.A. The author and publisher disclaim any liability in connection with the use or misuse of this information.

Acres U.S.A.
P.O. Box 1690
Greeley, Colorado 80632 U.S.A.
970-392-4464 • 800-355-5313
info@acresusa.com • www.acresusa.com

Printed in China

Originally published as Humussphäre: Humus—ein Stoff oder ein System by OLV Organisher Landbau Verlag, Im Kuckucksfeld 1, 47624 Kevelaer, Germany, phone +49 0 2832 97278 20,
info@olv-verlag.de, www.olv-verlag.de
© 2014, 2017 Organisher Landbau Verlag Kurt Walter Lau

Publisher's Cataloging-in-Publication Data
Names: Pommeresche, Herwig, author ; Lehmann, Paul, translator.
Title: Humusphere: Humus—A Substance or a Living System/ Herwig Pommeresche.
Description: Greeley, Colorado ; Acres U.S.A., [2019]
Identifiers: ISBN 9781601731487 (pbk.) | ISBN 9781601731494 (electronic)
Subjects: Farming—Organic | Soil management | Humus | Plants—Fertilizer | Plants—Nutrition.
Classification: LCC S605.5 .P6 2019 | DDC 631.4

CONTENTS

I. INTRODUCTION

Foreword xi
What Inspired Me to Write This Book xv

**II. GARDENING AND AGRICULTURE:
PAST, PRESENT—AND WHICH FUTURE?**

1. Agrobiology and Agricultural Chemistry:
   Two Sides of the Same Coin... 1
2. Suppressed Knowledge and Forgotten Models 39
3. The Significance of Microorganisms... 73
4. The Role of Water 91
5. Putting it to the Test: How Economical is Agriculture? 97
6. The Agriculture of the Future... 111
7. Organic Agriculture and Alternative Gardening 119

**III. GARDENING AND FARMING WITH THE HUMUSPHERE:
PUTTING IT INTO PRACTICE**

8. Working with the Humusphere, Not Against It:
   From the Stone Age through Today... 133
9. Practical Examples and "Recipes" 159

IV. FINAL THOUGHTS AND A LOOK TOWARD THE FUTURE

10. My Vision 189
11. Concrete Steps 195

V. APPENDIX.. 211

Glossary.. 213
Bibliography and Further Reading 233
Online Sources... 251
Index.. 252

About the Author

Herwig Pommeresche was born in Hamburg in 1938 and has lived in Norway since 1974. He received a degree in architecture from the University of Hanover. He has spent many years active as an architect and urban planner in Norway. After finishing his studies in architecture, he became a trained permaculture designer and teacher under the instruction of Professor Declan Kennedy.

Alongside other permaculture experts, he served as an organizer of the third International Permaculture Convergence in Scandinavia in 1993. He later served as a visiting lecturer at the University of Oslo. Today, Herwig Pommeresche is seen as a pillar of the Norwegian permaculture movement. He also serves as an author and a speaker.

He lives a partially self-sufficient lifestyle in accordance with the principles of permaculture and has kept himself busy over the course of the decades with traditional farm work, gardening, and keeping small livestock.

Herwig Pommeresche is a holder of the prestigious Francé Medal, awarded in 2010 by the Gesellschaft für Boden, Technik, Qualität (BTQ) e.V. (founded in 1993) in recognition of his contributions to organic methods and ways of thinking and to the preservation and improvement of the humusphere.

Section 1

Introduction

Foreword

*The greatest obstacle to discovery is not ignorance—
it is the illusion of knowledge.*
—DANIEL J. BOORSTIN, HISTORIAN, JURIST, AND AUTHOR

The supposed improvements to our food supply as a result of today's intensive agriculture methods are not what they seem to be at first glance, and they are only temporary in nature. And world hunger is not the only problem that remains unsolved. On the contrary, we also struggle with the "side effects" of industrial agriculture: Agricultural poisons have long been not only seeping into the groundwater and appearing in measurable quantities in the soil but have also found their way into human urine and breast milk. All around the world, we are suffering from a lack of nutrients in the soil and the loss of humic soil. And it's an undisputed fact that industrial agriculture has also contributed to global climate change. Much of the general public remains unaware of how seriously these negative effects impact our livelihoods.

Herwig Pommeresche has worked in gardening and agriculture for many decades, and in the process has thoroughly investigated a large amount of little-known research, both theoretically and in practice, and compared it with conventional methods. He has reached the following conclusion: the cause of the aforementioned problems is an inadequate understanding of the physiological and biochemical mechanisms through which plants absorb and process their nutrients. The current prevailing wisdom, in short, is that plants exclusively obtain their nutrients from salts dissolved within the water they absorb from the soil. This model, which the author calls the mineral model, has been the basis of the fertilization methods used in agriculture and gardening since the middle of the nineteenth century.

Starting from the beginning of the twentieth century, however, a body of research has emerged within biology and the other natural sciences that has investigated and described a very different type of plant nutrition: To put it simply, plants are able to envelop nutrient particles in their root cells and thereby transport the particles into the cells' interiors. This allows them to absorb not only very small material (such as ions or salts dissolved in water) but also larger molecules and even entire cells (microorganisms, for example); and they do so—and this is the most interesting part—in living form. This process is known as endocytosis, and the basic concept has been known within zoology and microbiology for a long time. What wasn't known for much of that time, however, was that higher plants also make use of endocytosis.

Little notice has been taken of this research in the natural and agricultural sciences, however, so the general public remains largely unaware of it (only very recently have there been a handful of studies that have taken another look at this phenomenon, and thus far they have confirmed it). Herwig Pommeresche has taken a critical look at our agricultural methods and analyzed and tested the practical applicability of the discoveries made by the researchers responsible for these studies. This second edition of *Humusphere* is the result of his own reflections and experiments and of those of the many kindred spirits who are also searching for sustainable alternatives in agriculture and gardening.

Admittedly, the explanations and theoretical bases for these observations may sound unlikely and hard to believe, and some aspects fundamentally contradict the things we all learned about plants in our biology classes. Even I found myself furrowing my brow when I read the manuscript for the first time. However, we should not dismiss out of hand the possibility that this "cycle of living material model" is indeed accurate. After all, life on Earth has regulated itself for the last 3.5 billion years according to principles that we—however much we might believe otherwise—are still far from fully understanding. About 300,000 years ago, this system of life gave rise to humankind, and it was around 10,000 years ago that our ancestors first began to engage in agriculture. But it wasn't until the modern intensification of agriculture that massive problems began to appear in our ecosystems and natural materials cycles, problems which had been unknown over the prior millions, or even billions, of years.

Is it really so far-fetched that the way nature governs and regulates itself on its own might work better than the methods we have come up with during this last blink of an eye in evolutionary history? Is it really out of the question that our methods are based on errors and faulty conclusions, considering that in fewer than two hundred years those methods have managed to destroy or bring out of balance great portions of the biosphere and our ecosystems, which had functioned on their own for millions of years before that? Any research that is truly conducted in the interests of the common good must impartially address these questions: Without us, how does nature secure and regulate plant nutrition? Why have our methods, in contrast, introduced so many problems within so few generations? Is it possible that we've overlooked something?

Any time that someone confronts us with new hypotheses and ways of thinking that we find unlikely, we should always keep in mind that what we call knowledge explains only a small proportion of our world, and it is always in flux. Even things that have long been taken for granted as true can nonetheless ultimately prove to be false. Just because many people consider something to be settled fact does not guarantee that they are not all wrong. I myself, just a few decades after my studies in biology, sometimes find that some piece of information that was taught to me at the time as irrefutably proven and established and thoroughly researched is being revised and presented much differently after again being looked into with newer methods and compared with other findings. How true might this be of a 170-year-old doctrine?

Today it is obvious to everyone that the Earth is round and not a disc and that it revolves around the sun rather than the other way around. We shake our heads at Galileo Galilei's contemporaries who threatened him with death simply because his discovery of the arrangement and movement of our solar system didn't fit within their framework. They were unable to conceptualize these ideas, whereas he was very much able to do so; but he could not prove them with the methods available to him at the time. He was nonetheless correct, as we know now. Who is to say, then, that all of our modern ideas and conclusions are correct? "We don't know what it is we don't know," is something Herwig Pommeresche frequently said to me while working on this book. How true. We should always operate on the assumption that we are still far from knowing everything.

Whenever we observe that something works differently—and indeed more effectively, more successfully, and without damaging side effects—with other methods than with the conventional ones, is it not logical to assume there must be some reason for this that we are not yet aware of? Why don't we set ourselves to the task of searching for this reason and expanding our knowledge? The clear practical successes generated by the methods presented in this book should give us plenty of reason to conduct further research in the direction suggested by the author in order to either confirm the model presented here or else find an alternative explanation for the good harvests that result from employing it.

It would be ideal if not only hobbyist gardeners and people looking for alternative methods for self-sufficiency take an active interest in this book's theories but also scientists and decision makers in educational and research institutions, as well as in politics. What do we have to lose by thoroughly researching this theory and investigating whether the methods can also be employed on a large scale in agriculture and food-crop cultivation? The negative side effects of our intensive agricultural methods are too serious, the outlook too dire, for us to continue to tolerate inaction. A solution must be found if our soil is to continue supplying us and future generations with enough healthy, toxin-free food.

In naturopathy—a field in which, much like agriculture and nutrition, differing opinions and ideologies often clash bitterly—there is a well-known saying: "He who heals is correct." During the work on this book, I came up with an analogous saying: "He who harvests is correct."

Eva Wagner
October 2016

What Inspired Me to Write This Book

INTRODUCTORY THOUGHTS

We still don't really understand what life as a whole really is, and we must not make the mistake of attempting to understand it using a model that is not suitable for explaining it.

This basic idea is what pushed me to write this book, and to this day it continues to be my motivation to search for gardening and agricultural methods that preserve this planet's plant, animal, and microbial forms of life rather than furthering their destruction—something that has been impossible to ignore or deny for a long time now. But thus far, holistic models based on the concepts of biological self-management and self-regulation have been de-emphasized in favor of purely technological concepts and models, resulting in the former's effective suppression. Our established method of "creating knowledge" does not permit the inclusion of the idea of self-regulating life cycles based on living matter in our general understanding of how the world works. In my opinion, our modern sciences are just not capable of viewing and treating life as something that cannot be fully comprehended using the established methods and models. We must therefore free ourselves from the false perception that the current state of the sciences is final and can no longer be changed.

The same goes for agriculture: our methods of thinking about and practicing agriculture represent only a small subset of all possible methods. So if it becomes apparent that this subset is completely insufficient or even based in falsehoods, with damaging or deadly effects on our biosphere, then we should examine it more closely. The biological aspects of agricultural research and practice are less comprehensively known than the chemical aspects. And among all of the books about modern agriculture that I have found, there is only one that questions the traditional understanding of plant nutrition: *The Organic Method Primer Update* by B. and G. Rateaver (Rateaver and Rateaver 1993). I want to emphasize that, astonishingly, this is true within both traditional and organic agriculture.

What I want to do now is assign the mineral theory (which forms the basis of our established methods of agriculture and gardening) its appropriate place, which is within the science of physical chemistry, a field that solely describes nonliving matter, and uncouple the theory from agriculture. This will move all of the other theories and discoveries that have been almost completely rejected or distorted in favor of the mineral theory to the forefront.

I look at any theory as a knowledge-creating explanatory model for phenomena and processes that we encounter or put into motion in nature. Based on my own numerous experiments, I believe that measuring the value of an explanatory model solely based on criteria set by the established sciences is an obsolete and outdated approach. I believe that you can also effectively measure such a model based on its practical success (i.e., empirically). This success must be defined in relation to tangible goals that you have set. Today's chemical-technological model has led to a system of agriculture that is poisoning the entire biosphere and thereby our food supply and us to varying degrees. The unchecked and now irreversible proliferation of genetically manipulated organisms, which have moved through the biosphere for a long time not only in the form of their pure chemical components but also autonomously and with "a will of their own," is another grave consequence. Some people see this as progress, but others have long recognized the catastrophic consequences of genetically modified agriculture.

THE CYCLE OF LIVING MATERIAL

It's safe to assume that there are very few traditionally taught plant physiology experts who are prepared to explore an entirely different explanatory model that either challenges what they've learned or calls it into question entirely. Because of this, new models are generally rejected at an early stage in the academic world, causing further research into them to be neglected or even prevented entirely. One such explanatory model—which can hardly be called new anymore—is the theory of the cycle of living material.

Today, this theory must be considered confirmed by the long-established phenomenon of plant endocytosis (also addressed in very recent studies; e.g., Samaj 2006; Paungfoo-Lonhienne et al. 2010, 2013; Kroymann 2010). The theory naturally leads one to look into historical cultivation methods such as the various forms of hydroculture used by historical cultures, which, as we can tell from reconstructions, achieved crop yields 5–15 times higher than those possible today with our chemical and technological processes, including genetic engineering. According to scientific studies, such as one carried out in Norway from 1929 through 1979, technological agriculture produces an increase in yields of 46 percent (NLVF-utredning 1980). However, attaining these crop yields requires the use of 400–500 percent more energy (according to the same sources), meaning that in reality this represents a *decrease*. This means that the alleged successes rely on abusing the topsoil. Recent attempts have also confirmed this, such as the experiences with winter rye.

All these figures are so far beyond the scope of modern practices, research efforts, and perceptions of agricultural concepts that they are never challenged—they are simply not acknowledged at all. I would like to use this book to push them back into prominence and to offer a possible explanation.

The transportation of living material through the biosphere provides a very different sort of plant nutrition than the absorption of pure salt ion fertilizers. The salt ions offer nothing besides their chemically defined characteristics. The proteins that all living organisms are built from, on the other hand, form very large molecules that—according to the current research—host a large amount of important organic information, but can also be chemically altered (which consequently alters their functions as well). And that is precisely why horizontal gene transfer (see pages 81 and 221) of manipulated proteins across species boundaries poses a danger with incalculable consequences.

We should also acquaint ourselves more closely with the theory that there are deep ecological relationships between plants, soil, and microorganisms, even though it appears to be so revolutionary (Crofoot 1989) that it has hardly been noticed or responded to so far and is accorded little value in the biological and agricultural sciences. In this book, I attempt to give organic agriculture a new and viable biological foundation based on the most recent discoveries.

The entire cycle of material in the biosphere can be described as "eating and being eaten." That phrase describes the most important concept in the cycle of living material. The conventional doctrine on plant nutrition within modern agriculture, on the other hand, is based entirely on the concept that the "dead" elements of the periodic table exclusively form the foundation of all living matter and thus of all plant nutrients. How living matter arises from these dead materials is not explained by the purely chemical model, however. The fundamental unit of the cycle of living material model, in contrast, is the smallest living cell components, whose metabolic pathways are traceable through the entire biosphere.

A CHANGE IN PERSPECTIVE: POINT OF VIEW IS THE KEY

As long as we limit ourselves to an explanatory model that looks at things exclusively from the point of view of chemistry, we will fail to fully understand organic processes or misunderstand them entirely. And that also throws our general understanding of biology—the study of living things—into question. As long as we cling to the pure mineral theory, we will continue to prevent the progress that is badly needed in agriculture.

Evidence and proof of the validity, the viability, and the believability of a theory—in our case, one of "alternative agriculture"—can be collected on many different levels. One way is internally (i.e., within the conventional scientific models that we all too readily accept as the only possible way to gain knowledge). But it can also be done externally, outside of these conventional methods. In our case, there is no other option but to be "unscientific" if we want to be able to present any alternatives for discussion at all. You might call it prelogical science.

Attempting a serious discussion without first establishing where everyone stands leads to senseless quarreling and will not end with a mutual understanding. Essentially, this means that you can only have a successful and comprehensible discussion of any given model or opinion if you start from the same standpoint and perspective. Otherwise, people—generally without consciously meaning to—just use the information they are presented with to strengthen and reinforce their own pre-existing views. But everything looks different when viewed from two different perspectives or points of view, and it is pointless to constantly argue about it. It's like the allegory of the four blind men who are taken to "see" an elephant for the first time. The first man stands at the elephant's head. He puts his hands on its trunk and says after a while, "Aha, an elephant is a long, soft tube that moves around—it reminds me of a snake." The second blind man stands underneath the elephant's stomach, between its legs, and contradicts the first: "But that's just not true. An elephant is made up of four thick, solid pillars!" Then the third blind man, who has only felt the elephant's tail, says, "You're both wrong. An elephant is nothing more than a long, thin whip!" And finally the fourth blind man, who is sitting on the back of the elephant and touching its ears, pipes up: "None of you has it right! An elephant is a large object that you can sit on, with two big pieces of canvas!"

As we can see, none of the four is correct. But none of them is entirely wrong either; each is accurately perceiving a portion of the elephant—but only the portion that he can understand from his own standpoint.

Herein lies the reason for the constant confusion whenever two people are discussing, say, organic carrots from two different points of view. And conversely, two different objects can seem the same when viewed from the same standpoint. This in turn is the reason why seemingly almost no difference is detectable between genetically modified and organically grown carrots. But this is not truly the case, as a series of experiments have shown that there is indeed a significant difference between their respective contents (e.g., Balzer-Graf 1991; Engqvist 1977; M. Hoffmann 1997; Pfeiffer 1958, 1984).

> **Perspective Influences Results**
> Which point of view something is being viewed, characterized, and finally "proven" from is of great significance. If, for example, you approach the problems in agriculture exclusively from the perspective of the salt ion model, you will come to different conclusions than someone who approaches it from a biological perspective.

In the ecological movement, there is a widespread perception that we are working with a different model from the "technological camp." And the perception is accurate. But upon closer examination, something very crucial is missing. Or, to put it better: the crucial thing has (without being noticed) been lost.

We must therefore be sure to carefully delineate different perspectives in situations where more than one is present. Thus far, this is rather uncommon in the sciences in general, including agriculture. Both standpoints originate in our reductionist Western scientific worldview, which already represents a limitation, even if we aren't aware of it.

Conventional agriculture is based on a chemical-mechanical model. Because of this, I refer to it as technological agriculture. Ecological agriculture claims to be based on biological concepts. After critical examination, however, I have concluded that people are no longer aware that biology also draws heavily on purely chemical-mechanical models itself while neglecting alternative perspectives. As long as this remains the case, ecological agriculture will never be able to establish itself as a truly distinct model for agriculture and nutrition.

NOTEWORTHY HARVEST YIELDS

The history and practice of agriculture and gardening is filled with results—measured either in yield per unit of growing area or by comparing the amount of energy employed in the process to the "harvested" energy—that can be described as surprising. But thus far, few of these cases have been documented or systematically evaluated. They are thought of as curious isolated cases, are little known, and are of essentially zero interest to the scientific community, because they cannot be explained or described within the bounds of our conventional models. Occasionally, a good gardening magazine will describe one of these cases along with planting tips, but these stories quickly fade into obscurity themselves: "Two kilograms of winter rye per square meter in the Sauerland." Or, "Eighteen kilograms of onions or twenty-two kilograms of carrots per square meter in Norway."

If yields on this scale are possible, how can it be that there are still so many starving people around the world? You can't eat money, but no money means no food in our system. World hunger is primarily a question of money rather than one of harvests or food. Join me on this journey and let's learn how it is possible for someone to get enough for four entire loaves of bread from just four grains of rye.

But the real goal really has nothing to do with the four rye grains or the four loaves of bread. My hope is to show the way toward an alternative agriculture, as I like to call it. The symbiotic-organic system of agriculture for which traditionalists and modern proponents of ecological agriculture have been striving for at least 170 years (since Justus von Liebig first applied the fundamentals of chemistry to agriculture), with varying levels of success, represents a wealth of new possibilities and opportunities.

THE GOAL OF THIS BOOK

In this book, I want to examine and systematize the research that lies at the core of the "life-emphasizing" model in order to create a new system of agriculture, and make this system available to researchers, farmers, and the consumers of agriculturally grown foodstuffs—which of course includes all of us.

I am aware that this model has many critics and opponents who consider it to be entirely wrong. I therefore ask you, dear reader, to draw your own conclusions: be self-sufficient not only in terms of your vegetables but in terms of your thoughts and opinions too!

Section 2

Gardening and Agriculture: Past, Present—and Which Future?

— CHAPTER ONE —

Agrobiology and Agricultural Chemistry

Two Sides of the Same Coin

THE NATURAL SCIENCES OVER TIME

Justus von Liebig: Chemistry and Agriculture

Justus von Liebig (1803–1873), the pioneer of today's predominant agricultural methods, was most interested in researching chemistry. His principal work, *Die Organische Chemie in ihrer Anwendung auf Agricultur und Physiologie* (*Organic Chemistry in Its Applications to Agriculture and Physiology*), was first published in 1840, with the ninth edition being published in 1876 (and the most recent reprint appearing in 1995), this being the edition cited in this book. In Liebig's view, the application and development of analytical chemistry was the most important thing of all, and he made agriculture fit into that framework. He did repeatedly assert that living things must be preserved, but his more recent successors have made sure that any conception of what is living is subordinate to the models used in chemistry. The tragic aspect of this is that the discipline of chemistry, by definition, can only describe nonliving material. According to this model, all living things are burned, dissolved, disassembled, mineralized, reduced to inorganic basic materials and their electrochemically described atomic compounds. For example, iron in all of its forms is reduced to the element Fe, nitrogen to N, phosphorus to P, potassium to K, and so on. The substances described by Liebig in his writings, which he referred to as both mineral substances and minerals as well as plant nutrients, are substances that would be called chemical elements today (i.e., building blocks). But according to Liebig, these building blocks are our nutrients in and of themselves.

But this way of thinking hasn't been tenable for a long time.

If you look at things from a purely physical perspective—that of classical atomic physics—nitrogen is and remains the same nitrogen regardless of whether it originates from plant or animal proteins. Because of this, people believed that it was possible to feed herbivores animal nitrogen (nitrogen coming from animal sources) as an "artificial diet"; for example, by using animal or fish meal as cow feed. Mad cow disease and other diseases have shown that this was a fallacy with grave consequences.

But according to Liebig's analytical methods, the ashes of these animals still contain the very same nitrogen, the most important nutrient to all organisms.

THE THEORY HAS HOLES

If you presuppose that all organic substances are being fully broken down to their simplest basic components, you exclude the possibility that qualitative characteristics might be passed along through a metabolic cycle. Applying this model makes it impossible to explain how, for example, diseases, resistances or immunities, the ability to adapt to climates, or the regeneration of vitality are passed on—and ultimately it is just assumed that such things do not happen at all. The mineralization theory thus rejects all the defining features of a plant's quality aside from those that are detectable through chemical analysis.

This is the reason for the belief that no visible or measurable differences can be found between organically cultivated soil or organically produced foodstuffs and their conventional (chemically grown) equivalents. It's not that the crops and their yields actually are identical or similar, it's that our research methods are inadequate to the task.

Or to put it another way: we don't know (yet) what we are and are not able to measure. As a result, we also don't know what it is that we don't know.

According to Liebig's own criticisms of older theories, a theory cannot simply be adapted in response to new ways of thinking and experiences. The old theory perpetuates itself, so to speak, until it reaches the point where its conclusions are clearly absurd:

Looking at the history of the natural sciences shows us that whenever a predominant doctrine is supplanted by a new one, the new one is not a further development of the old, but rather its direct opposite. An erroneous doctrine develops along the same lines as a correct one, but the one dies off because it does not have any roots while the other grows and is expanded upon. The erroneous doctrine, as it develops, ultimately leads to conclusions and viewpoints that are universally recognizable as irrational or even impossible, at which point it makes way for another that is its opposite; the truth is thus always the opposite of the error. (Liebig 1876, 9)

This brings up the question of to what extent the problems within agriculture at the beginning of the new century have reached the point of being "recognizable as irrational or even impossible by anyone":

The old doctrine assumed that the food eaten by plants was of an organic nature, that is, it was produced in the bodies of plants or animals. The new doctrine, on the other hand, supposed that the food eaten by green plants was of an inorganic nature, and that the minerals are converted within the body of the plant to carry out organic functions; the plant creates all of the components of its body from inorganic elements, forming the most highly complex compounds from basic building blocks, which form the basis of animal organisms. Due to its contrast to the earlier doctrine, the new one has received the name "mineral theory." (Liebig 1876, 9)

Liebig continues very clearly and unambiguously: "The food eaten by all green plants consists of inorganic or mineral substances."

Today, 170 years later, this claim is still perpetuated as universally true. And yet it has long been proven that plants also—perhaps even primarily—absorb nutrients another way. The prevailing conceptual model had already been confirmed to be false by the twentieth century, as proven by Artturi I. Virtanen in 1933 (Virtanen 1933; Virtanen 1958 in "Natürlich gärtnern" [*Gardening Naturally*], book 1).

The following was written in the 1990s: "On Earth, there are only two kinds of organisms with self-sufficient metabolisms: the sun-loving photoautotrophs (photosynthetic organisms), which use light as their energy source, and the rock-eating chemoautotrophs (also called chemolithotrophs), which do not even need light but only natural inorganic chemical reactions to derive their energy. Among all the organisms on the planet that might possibly have evolved such an independent mode of nutrition, only chemoautotrophic bacteria did" (Margulis and Sagan 1993, 129).

The many companies that have been producing effective microorganisms and living micronutrients for years must also be aware of this (more on this in the "Effective Microorganisms" section starting on page 84).

Since at least 1950, we've had a new theory—the theory of the cycle of living material—that is supported by strong enough evidence to supersede the so-called mineral theory. It postulates that living material ends up stored in the humus as the result of metabolic processes, at which point it is reabsorbed by plants as their food. Many attempts have been made, often desperately, all the way through to present day to reconcile the old theory with the new one. This has repeatedly led to confusion within ecology, with no results but uncertainty. In all likelihood, we will have to abandon the one theory in favor of the other. But who will have the courage to take the lead in doing so in the schools and universities?

AGROBIOLOGY AND AGROCHEMISTRY: A HISTORY OF THE LAST 170 YEARS

The search for life—for the components or elements that fundamentally define matter as living—is the true purpose of the field of biology. But in my view, biology has greatly neglected this task, with a pure chemical perspective coming to dominate the field. But there have always been important voices warning against this development while simultaneously researching alternatives and making their results available to the field. The following summary presents the most important of these that I have come across.

We begin with Liebig, who zealously spent his whole life working to apply chemistry to agriculture and plant physiology. At the same time, Charles Darwin was conducting biological investigations of the earthworm. Despite many significant objections, but perhaps unsurprisingly given Liebig's level of scientific ambition, the application of chemical aids won out over the usage of earthworms in the agricultural practices of the subsequent period. This is also expressed in a quotation from W. Hamm, written in 1872 in *Das Ganze der Landwirthschaft* (*The Whole of Agriculture*): "Then the great chemist Justus von Liebig appeared in the year 1840 and did away with the entire collection of older views on plant nutrition."

But also done away with were the ancient experiences collected and passed on by farmers over centuries, even millennia. The knowledge of and feeling for Mother Earth, for humus, and for the recently discovered life-forms living in the soil was almost completely lost. This was the beginning of farmers abandoning their knowledge and abilities in favor of trusting chemical corporations, resulting in farmers' knowledge and abilities being forgotten and suppressed over the course of what has now been 170 years as farmers are sent back again and again with a sack of artificial fertilizer to their ever more meager plots of land.

As a result, the history of human agriculture is always fundamentally described in terms of a battle for or against technological development. And biology has always been the loser of this battle.

The following is therefore an attempt to outline the history of agriculture from a biological perspective: the list gives the development of technological-industrial agriculture on the one hand and that of agrobiology and its understanding of plant nutrition on the other. This is only meant to serve as a basic collection of facts, however, without any claim to completeness and without being able to answer every question. This is because I consider it more important to show how and by whom the model of living material was and continues to be researched parallel to the model of dead material than to create a "complete" collection of facts.

1830-1870: Oliver Sheffield

Even before Charles Darwin and Justus von Liebig became famous, Oliver Sheffield was employing a new type of agriculture that made use of a systematic method of "earthworm husbandry" that allowed him to reduce his originally 3,000-acre field to 250 acres, yet still achieve higher yields. An interesting article in this context is "My Grandfather's Earthworm Farm," written by his grandson George Sheffield Oliver (1878–1968) and featured in *Die Wurzeln der Gesunden Welt, II* (*The Roots of the Healthy World, II*) by Wolfang von Haller as well as in the magazine *Natürlich Gärtnern & Anders Leben* (*Gardening Naturally and Living Differently*).

1837: Charles Darwin

In 1837, Charles Darwin (1809–1882) started investigating how the activity of worms contributes to the structure of field soil. But it wasn't until 1881 that he published his book *The Formation of Vegetable Mould through the Action of Worms*. By this point, Liebig had already spent his entire life intensively campaigning for his discipline, chemistry. The attempt to introduce "the earthworm in its applications to agriculture and physiology" to the field of agriculture never achieved the same level of success as chemical methods did.

1840: Justus von Liebig

Liebig (1803–1873) burned plants, expanded methods of chemical analysis, and isolated mineral components from the ashes of the plants. He led extremely ambitious efforts to apply discoveries from chemistry research to agriculture and plant physiology.

1848: Louis Pasteur

Pasteur (1822–1895) was active at the same time as Liebig. His life's work was on microorganisms. He discovered the anthrax and chicken cholera pathogens and developed a variety of vaccines.

However, this triggered an excessive systematic hunt for all manner of bacteria and viruses, which—much like the dogma of salt ions—has carried on to this day, 170 years later. I maintain, with good reason, that these two ideas are the primary causes of the worldwide poisoning of our biosphere. This misguided development is especially tragic when you consider the alternative possibilities that had been known and studied for a long time but were pushed out of the forefront as time went on.

1872: Wilhelm Hamm

In 1872, Wilhelm Hamm wrote the following about Albrecht Daniel Thaer (1752–1828), the publisher of the first comprehensive textbook on agriculture (Thaer volume 1809–1812): "He . . . assumed that plant food was made up of organic (combustible, arising from living beings, animals and plants) material found in the soil, and that the more of it was contained in a field in the proper state, the more fertile the land would be. This decaying material was known as humus . . . and it was believed that plants could absorb it on their own with use of the water contained in the soil" (4).

1875: Pierre Jacques Antoine Béchamp

The Frenchman Pierre Jacques Antoine Béchamp (1816–1908) published a significant work in 1883 under the title "Microzymas," in which he developed the idea that micrococci, which he called microzymas, were present in all living organs and that they played a significant role in animal and plant tissues. Béchamp believed that these symbiotic germs possessed a strong adaptive ability and that they could continue living beyond the life span of their host cells (Schanderl 1947). The theory of rhizobia in legumes was also being vigorously discussed by many authors at the end of the nineteenth century, and there was in fact not yet general agreement on whether bacteria originated from outside (as is accepted today) or from inside (i.e., from within the plant cell components; see Buhlert 1902; Hellriegel and Wilfarth 1888; Schultz-Lupitz 1895).

1876: Justus von Liebig

The year 1876 saw the publication of the ninth edition of Liebig's *Organic Chemistry in Its Applications to Agriculture and Physiology*: "It has proven to be the case [. . .] that the ash components are not just random components that vary from place to place, but rather that they play a role in the construction of the plant body; that these ash components thus play the same role for the plants that bread and meat do for humans or feed does for animals; that fertile soil contains many of these nutrients while infertile soil contains few; that infertile soil becomes fertile if you increase their quantities within it" (7). It goes on to say: "The old doctrine assumed that the food eaten by plants was of an organic nature. [. . .] The new doctrine, on the other hand, supposed that the food eaten by green plants was of an inorganic nature, and that the minerals are converted within the body of the plant to carry out organic functions; the plant creates all of the components of its body from inorganic elements [. . .]" (9). And it continues: "Due to its contrast to the earlier doctrine, the new one has received the name 'mineral theory.'" And: "The food eaten by all green plants consists of inorganic or mineral substances" (9).

The components of the ashes of burned plants are here explicitly declared to be nutrients, to the point of being compared with bread and meat, and Liebig is already unreflectively lumping together terms like "minerals," "inorganic elements," "mineral substances," and "mineral theory" here at the point where he first coined them. They haven't been critically scrutinized or differentiated since!

1876: Louis Pasteur

Pasteur proved that the interiors of plants are sterile. This has become accepted doctrine in the meantime, but evidence to the contrary exists as well (see Schanderl 1947 and Margulis 2001).

1898: Julius Hensel

Julius Hensel (1833–1903) published the book *Brot aus Steinen durch mineralische Düngung der Felder* (*Bread from Stones: A New and Rational System of Land Fertilization and Physical Regeneration*). Hensel addressed the subject of plant proteins with "explanations that were bizarre even for the time [. . .]" (Vogt 2000, 62). "Ammonia plants are weak, and so are the creatures that feed on them" (quoted in Vogt 2000, 63). Starting around 1890, Hensel established the "stone meal movement," which is considered its own distinct ecological agricultural system (Vogt 2000). "The chemical companies [. . .] launched [. . .] an expensive and rabid campaign to denigrate Hensel, keep his books out of print, and put a stop to his 'heretical' notions that NPK [nitrogen, phosphorous, and potassium] might be a toxin in the soil. [. . .] By the time the German chemical companies had amalgamated into the vast I. G. Farben conglomerate and brought their Führer to power, the last of Hensel's books was consigned to the flames" (Tompkins 1998, 182).

1911: Raoul H. Francé

In 1911, Francé (1874–1943) published *Das Edaphon* (*The Edaphon*), the first scientific summary and definition of the flora and fauna in the soil, and in 1921 he followed with *Das Leben im Ackerboden* (*The Life in the Soil*). (A new German edition that condenses the two volumes was published by OLV Verlag, Kevelaer, in 2012; see the bibliography starting on page 233.) *Die Letzte Chance* (*The Last Chance*), published in 1950, and *Humus*, published in 1957, later appeared in cooperation with Annie Francé-Harrar.

1913: Fritz Haber, Carl Bosch

In 1894 and 1899 respectively, Fritz Haber (1868–1934) and Carl Bosch (1874–1940) began to work on synthesizing ammonia. It was produced industrially by the chemical company BASF starting in 1913, but initially for bomb production during World War I rather than for increasing harvest yields (Smil 1997). It wasn't until 1920 that the first ammonia-based artificial fertilizer was manufactured.

1924: Rudolf Steiner

Concern over the loss of fertility in the living soil continued to mount, prompting Rudolf Steiner (1861–1925) to give his eight lectures on the "spiritual foundations for the renewal of agriculture" at the estate of Count Keyserlingk in Koberwitz, near Breslau. The goal of this agricultural course was to counteract the ongoing decline in fertility. Steiner took Goethe's phenomenological worldview a step further. He had a spiritually based conceptualization of agriculture and the natural sciences that also took microorganisms into consideration. The field of biodynamic agriculture, which was founded by Steiner, continues to be developed by various groups around the world. But the "hard" sciences reject this model. However, as the only holistic model originating in the Western cultural sphere, it is very significant in attempts at a "symbiosis" of agricultural models for a sustainable future (see also page 215).

1933: Artturi I. Virtanen

Virtanen (1895–1973) published experimental results that refuted the idea that plants feed on pure nitrogen, which had prevailed since the days of Liebig and Jean-Baptiste Boussingault:

On nitrogen nutrition in plants. [. . .] Through these decisive experiments, we have been able to prove that non-legumes can make exceptionally effective use of the organic nitrogen compounds that issue forth from the root nodules of legumes. It is clear that non-legumes, when growing together with legumes in nature or under cultivation, actually use these compounds directly, without first having to be broken down through the formation of ammonia and nitrates. [. . .] This finally proves that the conception of how plants consume nitrogen that has prevailed since Liebig and [Jean-Baptiste] Boussingault [editor's note: he carried out research on nitrogen enrichment in legumes] is not correct. The higher plants can do a very effective job of utilizing certain nitrogen compounds, including those that are actually present in the soil, without first having to break them down via the activity of ammonia-forming microorganisms. (Virtanen 1933)

The following years saw the beginning of the ongoing contamination of the biosphere with uncontrollable, artificially modified, synthetic chemical "dead material."

1940: Albert Howard

In 1940, the British botanist, mycologist, and pioneer of ecological agriculture Sir Albert Howard (1873–1947) described mycorrhiza symbiosis, in which plant roots are able to absorb carbohydrates and proteins. The "fungal root" thus forms a living bridge between plant and soil (Howard 1940).

1947: Hugo Schanderl

Schanderl (1901–1975) published his most important book, *Botanische Bakteriologie und Stickstoffhaushalt der Pflanzen auf neuer Grundlage* (*A New Basis for Botanical Bacteriology and Plant Nitrogen Balance*) in 1947. It contains a comprehensive description of the history of bacteriology, which I've quoted from here. He gives instructions on how to reproduce broad-ranging experiments that show how mitochondria and chloroplasts (important cell components, or organelles, that have their own DNA) "remutate" from dying cell structures into autonomous, living, multiplying bacteria. Schanderl further refined the cycle of living material model and a few years later—from 1964 through 1970—attempted once again to find a receptive audience for his research, challenging the field of bacteriology to move away from monomorphic thinking and toward an evolutionary mind-set.

"Modern bacteriology is still too firmly rooted in the monomorphism of the previous century. It still hasn't begun to think evolutionarily" (Lovelock 1991; Margulis and Sagan 1993; see also "The Gaia Hypothesis" on page 65). "Over the course of evolutionary history [. . .] primordial cells [. . .] came together into larger cooperative units. The constituent members of these larger cells lost their independence in the union, and were assigned special roles in the new larger cells, eventually becoming functional organs and organelles" (Margulis and Sagan 1993). "If some outside influence disturbs or destroys this organization, the constituent units once again take on independent lives and return to their previous low level of organization" (quoted from *Boden und Gesundheit, Zeitschrift für angewandte Ökologie* [*Soil and Health, Magazine for Applied Ecology*], no. 45 (1964), pages 10-12; no. 66 (1970), pages 7-10; no. 68 (1970), pages 7-10; see also "Remutation" in Schanderl 1947).

1950: Annie Francé-Harrar

The 1950 book *Die letzte Chance für eine Zukunft ohne Not* (*The Last Chance for a Future Without Hardship*) by Annie Francé-Harrar (1886–1971) is probably the only summary of the scientific history of the worldwide humus catastrophe (see also page 236 and the following pages). Francé-Harrar's central theme is that the ongoing decline in humus content caused by ten thousand years of human agricultural activity is responsible for entire cultures dying out, people's migrations, colonialism, and world trade. The same sort of reasoning can easily be applied to current circumstances in the European Union and to certain aspects of globalization. Her writings suggest that decentralization probably represents the only hope for new humus production.

1950: Mineral Fertilizers Come to Dominate around the Globe

This process began at around the same time (i.e., after World War II).

1950–1980: The So-Called Green Revolution

The green revolution turned into an economic and chemical war on as yet nonindustrialized agriculture around the world. At the same time, the "genetic revolution" was taking hold, with genetic modification being the final violent, technological attack on living material (so far). This initiated the ongoing contamination of our biosphere with technologically manipulated living substances that propagate themselves beyond our control.

1955: Hans Peter Rusch

With the help, as always, of his predecessors, Rusch (1906–1977) embarked on an extensive speaking tour of nine European cities between 1949 and 1953, and in 1955 published a collection of these lectures under the title *Naturwissenschaft von Morgen: Vorlesungen über Erhaltung und Kreislauf lebendiger Substanz* (*The Natural Sciences of Tomorrow: Lectures on Conservation and the Cycle of Living Material*).  This represented the beginning of the second great attempt to establish the principle of living material as the foundation of biology and agriculture.

In 1968, he summarized his work with the publication of *Bodenfruchtbarkeit: Eine Studie biologischen Denkens* (*Soil Fertility: A Study of Biological Thinking*; Rusch 1968; new edition 2014). Together with Hans Müller, he made efforts to put organic agriculture into practice. These efforts continue today, spearheaded by Orbio in

Scandinavia and the Bioland-Verband für Organisch-Biologischen Landbau e.V. in Germany, but neither movement has been able to establish its most important principle, namely the cycle of living material, either as a common term or as a foundation for further research. In *Kultur und Politik (Culture and Politics)*, Rusch wrote:

"Mineralization" of living material. [. . .] At the beginning of this century, scientists were in agreement [. . .], with few exceptions, that this formulation was accurate: "any organic material must be 'mineralized' in the soil before plants can absorb it." The view was that no organism, not even animals or humans, was capable of absorbing large organic molecules into its body. [. . .] Much has changed since then. [. . .] Anyone who still talks about a complete breakdown of all nutrient substances into mineral salts today has either slept through this entire eventful period, or is pursuing particular economic aims that no longer bear any relation to science. The discovery of vitamins and enzymes, which was around four or five decades ago already, proved that the nutrient cycle also includes larger compounds that can indeed be absorbed by humans, animals, and plants and that move through their metabolisms without issue. But that was just the very beginning: today it is absolutely certain that every organism is able to absorb large molecules of living material, including entire bacteria even, when they are available as nutrients. (1974, 12–17)

1958: Nikolai Aleksandrovich Krasilnikov

Krasilnikov (1896–1973) published a paper that suggested that plants seem to feed not only autotrophically, but heterotrophically as well (Krasilnikov 1958, 1961).

1962: Publication of Silent Spring

Rachel Carson's *Silent Spring* became known to a broad audience. The book gives a sobering description of the destructive consequences of the worldwide use of pesticides and herbicides.

About the Same Time: Beginning of Organic Agriculture

Following the principles espoused by Müller and Rusch, the following years saw the start of alternative agriculture, thanks to the recognition that plants feed on proteins and living cells.

1968: Publication of Bodenfruchtbarkeit: Eine Studie biologischen Denkens

Soil Fertility: A Study of Biological Thought by Hans Peter Rusch is still in publication today (with the latest edition released in 2014). However, the model it advocated, with the cycle of living material as the basis of plant nutrition and agriculture, has still made no noteworthy headway into our agricultural and gardening practices.

1970: Lynn Margulis

Lynn Margulis (1938–2011) was already demonstrating the necessity of thinking in evolutionary terms in 1967 when she published her serial endosymbiotic theory. Margulis believed that primordial bacteria were the first form of true life and that all further forms developed from them. This provides a very modern model for understanding the results from Francé, Steiner, Schanderl, and Rusch that we know from the previous century—or have forgotten about entirely. During the following thirty years, she fought to win acceptance for this model, much like Schanderl thirty years before her. Around 1970, as Schanderl and Rusch's careers as working scientists were coming to a close, she began her thirty-year career as a biologist and continued the fight for living material against the same scientific conformity that her predecessors had been confronted with while pursuing the same goal. Although she probably was completely unaware of Schanderl, the two had an important source in common. In 1947, Schanderl was already citing the American biologist Ivan Emmanuel Wallin (1883–1969), from, among other sources, his work *The Independent Growth of*

Mitochondria in Culture Media. Margulis also cited from Wallin's *Symbionticism and the Origin of the Species* (1927).

In 1998 she wrote the following about Wallin in her book *Symbiotic Planet* about Wallin: "One of my most important scientific predecessors thoroughly understood and explained the role of symbiosis in evolution." She goes on to write: "That animal and plant cells originated through symbiosis is no longer controversial. Molecular biology, including gene sequencing, has vindicated this aspect of my theory of cell symbiosis" (15).

1972: The Club of Rome

In 1972, the members of the Club of Rome wrote: "The average per capita amount of agricultural land is 4,000 square meters for the world as a whole. In the USA, there are 9,000 square meters per inhabitant, and 2,000 in Switzerland" (Hitschfeld 1995).

1978: Bill Mollison and David Holmgren publish Permaculture One

The two founders of the permaculture movement ("permaculture" is derived from "permanent agriculture"), which has become a global phenomenon, have produced many writings and offered courses around the world to educate nonspecialists on the paradigm shift that has been brewing in Western science for a hundred years now.

1980: Report of the NLVF-utredning (Norwegian Institute for Agricultural Research)

Under the heading "Energy Usage for Food Production in Norwegian Agriculture, 1929–1979," the NLVF-utredning report states: "Production has increased by a factor of 1.5. Energy expenditure has increased by a factor of 4.5" (1980; see also "Ökologie & Landbau" ["Ecology and Agriculture"] 2000).

1982: Fritjof Capra (Born 1939) Writes about Organelles

At an even smaller scale, symbiosis takes place within the cells of all higher organisms and is crucial to the organization of cellular activities. Most cells contain a number of organelles, which perform specific functions and until recently were thought to be molecular structures built by the cell. But it now appears that some organelles are organisms in their own right. The mitochondria, for example, which are often called the powerhouses of the cell because they fuel almost all cellular energy systems, contain their own genetic material and can replicate independently of the replication of the cell. They are permanent residents in all higher organisms, passed on from generation to generation and living in intimate symbiosis within each cell. Similarly, the chloroplasts of green plants which contain the chlorophyll and the apparatus for photosynthesis, are independent, self-replicating inhabitants in the plants' cells. (Capra 1994, 308)

1988: James Lovelock Publishes The Ages of Gaia

Its German translation was published in 1991 under the title *Das Gaia-Prinzip*.

1989: Explanation of Plant Endocytosis

Peter Tompkins wrote about John Hamaker and Bargyla Rateaver and explained plant endocytosis.

1993: Further Research into Plant Endocytosis

B. and G. Rateavers' *The Organic Method Primer Update* was published with a collection of images and writing documenting plant endocytosis (Rateaver and Rateaver 1993, 1994).

1993: Dorion Sagan and Lynn Margulis

In 1993, Sagan and Margulis published *Garden of Microbial Delights: A Practical Guide to the Subvisible World* (Margulis and Sagan 1993). The work comprehensively covers the subject of heterotrophic plant nutrition.

1997: Vaclav Smil Writes About Agricultural Efficiency

In *Cycles of Life*, Vaclav Smil explained how traditional agricultural methods from China, Egypt, and other parts of the globe were able to feed more people per hectare than, for example, the methods used in North America in the nineteenth century (Smil 1997).

1997: Louise E. Jackson on Nitrogen

In *Ecology in Agriculture*, the American agroecologist Louise E. Jackson wrote about nitrogen nutrition in plants and demonstrated that the mechanisms by which plants absorb such nutrients has not by any means been fully explained (Jackson 1997).

1999: Lynn Margulis Publishes Symbiotic Planet

All her life, the renowned biologist demonstrated that she had the courage to swim against the scientific current. Her endosymbiotic theory, according to which all animal and plant cells at some point formed via the amalgamation of various types of bacteria, met with heavy resistance for many years (see Schanderl 1947, 1966, 1970; and Rusch 1953–1968), but now it is accepted as standard doctrine.

1999: Günter Klaus-Joachim Blobel

Blobel received the Nobel Prize for medicine for his discovery that "proteins have built-in signals that govern their movement and their location within the cell."

1999: Awards for Medicine Are Given in Norway

Awards bestowed included recognition for research within evolutionary biology (Irma Thesleff, Helsinki, Finland); for the discovery of new receptors for the transportation of biologically important molecules and molecule complexes within cells (Søren Kragh Moestrup, Aarhus, Denmark); for the identification of several molecules that regulate membrane transport (i.e., endocytosis); and for the discovery of new principles by which proteins regulate important cellular functions (Harald Stenmark, Oslo, Norway).

2000: Anne Holl and Elisabeth Meyer-Renschhausen publish Die Wiederkehr der Gärten: Kleinlandwirtschaft im Zeitalter der Globalisierung (The Resurgence of the Garden: Small-Scale Agriculture in the Age of Globalization)

This book showed that in spite of—or perhaps as a result of—globalization, we are experiencing a resurgence in gardening. More and more people are turning to small-scale growing. Between then and now, this movement has developed rapidly around the world, in part in the form of "urban gardening" (see also pages 107, 150 and 154).

2000: Gunther Vogt

Despite an admirable level of familiarity with the source materials, Vogt portrayed Virtanen, Schanderl, and Rusch as unreliable in his dissertation (very much along the lines of reasoning of the conventional model) with the help of a wide variety of quotations. He even described Rusch as "lacking in scientific respectability" (Vogt 2000). This is how these researchers are "culled" from the established sciences.

2002: Peter Schneider

Schneider published *G. Enderleins Forschung aus heutiger Sicht* (*G. Enderlein's Research from a Modern Perspective*). It discussed Béchamp, Schanderl, and Margulis in context (Schneider 2002).

2006: Elke Krämer

In 2006, Krämer published the work *Leben und Werk von Professor Enderlein (1872–1968)* (*Life and Work of Professor Enderlein [1872–1968]*), which served as her dissertation in medical history. During World War I, Günther Enderlein discovered that certain microbes and the overall milieu of the body are responsible for our health. This was based on dark-field live blood analysis.

Twenty-First Century: Stem Cell Research

Stem cell research has taken on great importance during this century. It is concerned with cells that have the ability to develop into a wide variety of different types of tissues and organs—a major field of research in medicine, biology, and related disciplines.

The question suggests itself: Is it Hugo Schanderl's remutated organelles from 1947 through 1970 and Hans Peter Rusch's regenerating cells from the cycle of living material that are being celebrated by geneticists as "wonder cells that can repair anything" here? Changes are afoot, and it would be irresponsible for organic agriculture not to react.

> **What Has Become of the Study of Life?**
>
> After assembling this long list, I asked myself: Where is research on the subject of life to be found now? My answer: in the Gaia theory and its proponents, James Lovelock and Lynn Margulis (whom I come back to on page 65). Agriculture, including organic agriculture, can and must draw on this theory and continue researching it, even if it means temporarily moving away from traditional established science.

AGRICULTURE AND CHEMISTRY
Whom Does Agriculture Serve?

I would like to pose an intentionally provocative question: Who should determine the future of agriculture and thus our food supply? Should it be the field of chemistry or the field of biology?

Biology needs to concentrate more strongly on the living processes within organisms and look at them as just as important as the fundamentals of chemistry, and agricultural science and biology need to work much more closely. My appeal is: This new biology is obliged to support agriculture in introducing new explanatory models. Furthermore, the difference must be explained to consumers of the products in a comprehensible way.

Some may say that this is impossible, but that is true only of our conventional view of chemistry. By its own definition, chemistry cannot describe life and living systems and its explanatory models cannot cover them. The difference between the two models or methods, however, is significant and verifiable. If the ecological scene is not ready embrace this, I am convinced we will be unable to solve the urgent problems facing us today within agriculture. I believe that it is absolutely necessary to get a new, synergistic biological movement going that is clearly and explicitly based on the fundamental concept of living material. This will allow us to establish a confident "biological" biology as well as a confident biological agriculture. The idea of the cycle of living material has been challenged and "scientifically" rejected many times by proponents of the current model. But in the last thirty or forty years, molecular biology has become so independent that outsiders to it—such as Margulis and the supporters of the Gaia theory—have once again turned to the idea of the theory of living material. Recent research in molecular biology (e.g., Kroymann 2010) has made it possible to develop and provide support for older hypotheses and models of the cycle of living material. There is a good chance, with the help of the endosymbiotic theory and the Gaia theory, of constructing a believable, viable model that finally brings together the "science of tomorrow" (Rusch 1955; Schanderl 1970), thereby making a broader audience aware of our new understanding of agriculture and plant cultivation. An important result would be that the practically senseless application of more and more outside energy and materials in agriculture would come to an end, and their places would be retaken by the care, nourishment, and propagation of the life in the soil.

To accomplish this, the biological aspect of this problem needs to be, if not fully separated, then clearly and comprehensibly delineated from models originating in the field of chemistry, because metabolic processes—and here in particular plant nutrition—encompass a large field of research.

W. Hamm wrote on this subject in 1872 about Albrecht Thaer:

He, and with him every farmer, *assumed that plant food was made up of organic (combustible, arising from living beings, animals and plants) material found in the soil, and that the more of it was contained in a field in the proper state, the more fertile the land would be. This decaying material was known as humus . . . and it was believed that plants could absorb it on their own with use of the water contained in the soil. But this was an error. Some researchers had already . . . shown that plants acquire the carbon that they need from the air, and soil or mineral material was already recognized as a component of their bodies, and some doubt had arisen concerning the earlier doctrine.* Then the great chemist Justus von Liebig appeared in the year 1840 and did away with the entire collection of older views on plant nutrition. He proved that a number of mineral components—potash, phosphoric acid, lime, iron, clay, magnesium oxide, sodium bicarbonate, silica, sulfur—make up the fundamental nutrients of plants in the soil; they can be recovered from the ashes of burnt plants. (Hamm 1872)

I am going to attempt to show that Thaer's assumption might not have been an error after all.

THE ETERNAL SEARCH FOR NITROGEN

The biosphere contains a certain quantity of nitrogen in the form of organic (and generally living) compounds. According to Frederik Vester (1987), 200 billion tons of organic material are converted on and in the soil each year—as we shall see later, primarily through "eating and being eaten" processes in the form of the metabolism of living material. This vast amount of organic material is the "waste" generated by the life processes that keep the biosphere running in all its variety. But it is also the nutrients for the following year's life processes! In this "waste," much of which is living (1–2 tons of living organisms per 1,000 square meters of uncontaminated agricultural soil), the much-coveted nitrogen is relatively firmly bound, by, for example, being built into the organic structures of protein molecules (among other substances). This means that it cannot be washed out by the large quantities of water that are constantly flowing around it. Relevant to this issue are the findings of Virtanen, Schanderl, and Rusch, who showed that organic substances do not need to be reduced to inorganic ions in order to be absorbed by plants as nutrients once again.

Now let's imagine an equal quantity of the synthetic, easily accessible nitrogen that we have put into circulation in the biosphere. For one thing, we have a very imprecise idea of what actually lives in the soil. Furthermore, we have essentially zero knowledge about which of the life-forms that aren't known to us perish when we regularly apply large amounts of synthetically produced nitrogen salts.

On this subject, I will cite Gerhardt Preuschen from his *Ackerbaulehre nach ökologischen Gesetzen—Das Handbuch für die neue Landwirtschaft* (*Agriculture According to Ecological Laws—The Handbook for the New Agriculture*), written in 1991:

"For as long as people have believed that plants can subsist exclusively on water-soluble substances, they have understandably attempted to find these substances or compounds in the soil and to check the plants' nutrient supply against an established amount, usually in a water-soluble state, and to classify them as signs of fertility. We know today that this theory was incorrect. Under very adverse conditions, plants can process water-soluble material, but they always need microbes to do so. This whole system of direct transfer of material into the plant's body leads to diseases while at the same time damaging the life in the soil. To put it briefly—the entire mineral theory and the way that it has been applied was the wrong approach. We can also proceed on the assumption that the data used to determine actual fertility is unrelated and thus uninteresting, and in fact must sometimes be analyzed the opposite way (69)."

And he continues: "Free nitrates practically never appear in an undisturbed ecosystem. If this were not the case, rivers would have to have been carrying nitrates for centuries, and mountains of sediment would have formed in the seas and oceans, or they would contain some nitrate content. It is astonishing that scientists who want to be taken seriously continue to repeat the claim that nitrates are a natural component of the living soil and an important plant nutrient" (143).

Plants' nitrogen supply is precisely regulated in nature. Because this aspect of plant nutrition is being ignored, we have an excess of easily soluble nitrogen compounds today.

In *Ecology in Agriculture* (Jackson 1997), we find the following statement on this topic in the article "Nitrogen as a Limiting Factor": "Unknown are the mechanisms through which crops acquire nitrogen from the soil" (163). This means that it is not known to science at all how plants absorb and metabolize "mineral" nitrogen fertilizers (or other forms of nitrogen).

The nitrogen cycle in the biosphere is out of balance. Peter Vitousek from Stanford University ("World Watch," Sept/Oct 1997) and Vaclav Smil (1997) point out that we—through the production of artificial fertilizers and by burning fossil fuels—have already produced and released as much easily soluble nitrogen as is contained in an inert form (i.e., chemically stable, not readily releasable) in organic material in nature. This causes at least as many problems as the excessive release of carbon dioxide and its well-known effects (the greenhouse effect and climate change). We are only just beginning to really think about this issue.

In contrast to manure and other organic fertilizers, which pass along already-bound nitrogen from one organism to another via the cycle of eating and being eaten and thereby keep the overall cycle of materials (including the purification process) in running order, the release of millions of tons of synthetically produced and highly reactive nitrogen each year represents a new problem whose scope has barely been anticipated. If the biological need of all organisms to take in biologically integrated nitrogen (e.g., in proteins) from organic substances (by eating) is neglected in favor of water-soluble, easily reactive nitrogen salts, the entire highly complex cycle of material in the biosphere may be endangered. If it were to collapse, we would very quickly find ourselves drowning in our own proverbial filth.

The fact that intensive agriculture also contributes to climate change by releasing large amounts of nitrogen in the form of nitrous oxide, which is a greenhouse gas like carbon dioxide, is yet another issue. The problems associated with the chemical contamination of our water sources and groundwater by water-soluble nitrogen compounds have also been known for a long time.

We must assume that these high salt concentrations not only decimate the life in the soil but also that they strongly affect plants' root mucilage and the microorganisms living within the soil. If you only make easily accessible salt ions dissolved in their "drinking water" available to the plant roots (which it doesn't contain naturally, at least not in such high concentrations; Preuschen 1991), is it not conceivable that this can disrupt the entire collective metabolic process of the biosphere? Does the assumption not suggest itself that the metabolic processes, in their life-essential entirety, may be bypassed by this excessive contamination with easily reactive nitrogen? It is the self-regulated constraint on nitrogen availability itself that creates the proper conditions for the fully contaminant-free cycle of living and dead material. The term "humus sapiens" (i.e., "intelligent humus") occurred to me to describe this: the soil already knows what it needs to do.

WHAT HYDROCULTURE REALLY IS

The constantly repeated claim that plants can only make use of water-soluble salt ions as food can today be accurately viewed as obsolete and even as incorrect. You can find countless evidence in books that only certain very simply structured microorganisms can subsist in an entirely photo-chemoautotrophic manner. All other forms of life are forced to "eat others," meaning that they are all likewise also heterotrophic. Questioning experts on whether they can name a purely chemical form of plant nutrition prompts them to bring up hydrocultural systems that work on a "purely" chemical basis. But there is almost nowhere that you can find more bacteria and algae than in these "artificial" salt solutions! To put it another way: practically no viable plant hydrocultural systems are free of life-forms such as plankton and other microorganisms (see pages 91–96). And if you want to work under absolutely sterile conditions—which is only possible in a laboratory, if at all—you can refer to Hugo Schanderl about how new microorganisms are capable of reproducing from living cells, or more precisely, from their mitochondria and chloroplasts, a process that we will later learn about under the name of remutation (see page 43). Even a theoretical germ-free environment will be settled by these microorganisms.

Plants are therefore—with few exceptions—not pure primary producers. They do work partially photo-autotrophically and also chemo-autotrophically above ground. But underground, they live heterotrophically. There their root hair cells, together with the root mucilage and the microorganisms living within it, eat anything that they can reach: salt ions, if they appear in originally pure "drinking" water, true minerals in the form of stone dust particles, amino acids and entire large protein molecules, and also bacteria and other microorganisms (this process, called endocytosis, is explained on page 39).

"They eat you in the end" (Mollison 1989). This is reminiscent once again of the Bible verse, which in its Norwegian form is: *Fra jord er du kamen, til jord skall du bli, og fra jord skall du gjennoppstä* ("from earth you came, earth you shall become, from earth you shall be resurrected").

THE ONGOING POISONING OF OUR SOIL AND WATER

Anything that is put in during a production process must of course come back out again, either in the form of delivered goods or as solid waste, exhaust gas, sewage, or more recently through the profit-friendly method, misunderstood as the environmentally friendly method, of recycling resources (in the wrong way) into other products. This just makes the waste creation process somewhat longer, but you're essentially just shifting the problem around. Over the last decades, approximately 70,000–100,000 synthetic substances have been brought into the world; around 10,000 substances are added each year. Once they've been unleashed, these materials are neither controllable nor recallable and they all end up somewhere in our biosphere sooner or later. After fifty years (Carson 1962), there really shouldn't be anyone who hasn't heard of dichlorodiphenyltrichloroethane (DDT) and polychlorinated biphenyl (PCB) being in the blood of polar bears and human breast milk.

In potato cultivation—to name just one concrete example—there are a number of different pesticides that can legally be used. Their instructions for use include information on how many days it takes for the harmful substances to break down. This leads to the mistaken idea that the pesticides will no longer exist after this point. There may even be many scientists who are convinced of this themselves. But it is not the case! Every single chemical compound follows its own path through the sky, the soil, the water, through humans and animals—and through potatoes. In addition, even if the poisons are applied individually, they still end up finding each other out in the wild. At that point, any substance can encounter any other, even in the first "generation," and even at this first meeting they can form new chemical compounds. From one step to the next, these compounds naturally carry their arbitrary "acquaintances" along with them. Assuming 70,000 synthetic substances, a simple calculation tells us that there are 2.5 billion possibilities. What actually happens cannot be determined from any calculation. But the chemical catastrophe goes further still as the molecules of the artificial substances are broken down into segments. These also "multiply" in accordance with the same statistical rules just mentioned.

This produces a chain reaction of chemical substances, whether they are natural or synthetic, that is neither predictable nor controllable.

Many of the artificial chemical compounds we produce also come with an additional ecological problem: many of these substances are produced under high pressure, high temperature, and/or other unnatural technological conditions. This is exacerbated by energy being employed to speed up the production process (many chemical reactions proceed more quickly at high temperatures). This means we are constantly developing substances that nature takes a very long time to incorporate into its metabolic cycle—if it can do so at all. When it does, living cells and organisms store the foreign substances (which can also be toxic) in places where they can do the least damage or where it is easiest to accumulate them. For example, DDT and PCB collect in the fatty tissue of animals, where they are first released when mammals form energy-rich, fatty milk for their offspring in their bodies. So the first things we offer our children are DDT and PCB.

And we haven't even addressed the metabolites, the substances that have formed and continue to form "in the wild" through the breakdown of and chemical reactions between all these chemicals we've released into nature. There has hardly been any research on them anywhere in the world yet.

We therefore see that chemistry is incapable of controlling "the spirits it has conjured" (to paraphrase Goethe). Worse still is the possibility of deliberate abuse.

BOUND CONTAMINANTS

Are Laboratories Overlooking Toxins in the Environment and the Food Supply?

Food chemists have a problem: their analyses often claim significantly lower residual contamination than actually exists. What I am referring to are "bound residues" (that is the harmless-sounding internal designation, at least). Officially, experts happily reassure us with the statement that bound residues in foodstuffs pose only a limited risk. Pesticides or pharmaceuticals are, according to information from the producers, quickly broken down or excreted. Countless analyses have confirmed that the amount of these substances present sharply declines after their application. But skepticism arises nonetheless when one is aware that, for example, routine analysis of the antibiotic chloramphenicol shows that only 0.2 percent of the administered dosage is still detectable. Where might the rest be?

The explanation is simple: the common methods of analysis predominantly only capture the unaltered initial substance. And here lies the crucial point: all life-forms attempt to rid themselves of dangerous substances as quickly as possible. Both humans and animals join these substances with a "transport vehicle" in order to make them water soluble. Then they can excrete the undesirable substance through their urine. Plants, on the other hand, have to store them somewhere within their cells. They make the toxic substances incapable of causing damage by binding them to cell components, keeping the substances' acute toxic effects from harming the plants.

This natural process means the substances are indeed no longer detectable—but they haven't disappeared by a long shot.

These facts also throw the legally prescribed time period between when a pesticide is applied (or a drug is administered) and when the product can go to market into question. By the end of this period, the harmful substances will have been broken down beneath the level where they can be detected or will be detectable only in "harmless" quantities. But if no more *free residues* can be detected, that generally just means that that they are now present in the form of *bound residues*.

The question is: why don't laboratories also analyze the bound residues?

In practice, this represents an almost insurmountable task for a chemist, because the possible ways for residues to bind with other molecules or cell components are, as one might imagine, numerous; plants, like animals, can bind contaminants to a wide variety of cell components. So you would have to proceed differently for practically every food and every substance. There have been a few modern detection methods developed recently that work via immunochemical reactions. But they are presumably still too expensive for routine analyses.

Even if the bound residues represent a nearly impossible task when it comes to routine analysis, it has been known for years that their remnants can be easily tracked by radioactively marking pesticides. Tests of this type have shown that foodstuffs generally contain many times as much of the residues as chemists can find using conventional methods. For example, radishes were found to contain 100–1,000 times as much of the pesticides dieldrin, permethrin, and carbofuran as was falsely shown through routine analysis. Similar results have also been attained for many pesticides in every imaginable foodstuff, including organophosphate toxins like malathion and pirimiphos-methyl and organochloride insecticides like DDT.

The official data on residues from food-safety authorities is thus incomplete. However, this knowledge generally only gets out through chance. So the academic world was thoroughly confused when significant traces of the stalk shortener chlorocholine chloride (CCC)—also known by its brand name, Cycocel—were found in portobello mushrooms, because stalk shorteners are used only in grain cultivation. The CCC came from the straw that the mushrooms were bred on. Almost none could be found in the straw via analysis. But the mushrooms broke down the straw as they grew, which freed the CCC again. And they absorbed it, leading chemists to discover it again in the mushroom tissue.

Studies on bioavailability and toxicological experiments have shown that the bound residues are much more important than the free residues (which thus far have always been what was measured) in determining maximum allowable quantities. Accordingly, feeding experiments with "healthy" foods have produced undesired effects, like changes in the blood count (e.g., reductions in white blood cells, a sign of a compromised immune system), as well as changes to the messenger substances in the brain.

Consumers generally hear nothing of the existence of bound residues. A further example of this is that the pesticides deltamethrin and fenvalerate, in the form of bound residues in contaminated grains, have been broken down by lab animals in their digestive tracts. Valiant researchers have warned: "For a proper assessment of the safety of pesticides to the consumer, information about the formation of bound residues is essential." And "bound residues must be included when determining maximum allowable quantities of pesticides and when evaluating their toxicity." But this type of risk is not discussed publicly.

Everything Organic, or What?

The preceding information is also significant in assessing organic agriculture. Experts have often claimed that as long as the minimum required time between spraying the pesticide and the food product hitting the market is adhered to, there is barely any detectable difference between organically and conventionally grown food. This conclusion is not surprising given that the bound residues are not taken into account due to laboratories being unable to measure them in routine analyses. According to their methods, the pesticide content is too low to be detected and conventionally grown products seemingly present hardly any difference from organic vegetables.

What happens to harmful substances like pharmaceuticals that are excreted from animal or human bodies? The liver makes them water soluble by, for example, joining them with certain sugar substances. These "joined" pharmaceuticals can then be expelled through urine. They will have lost their pharmaceutical effect. But in contrast to the view of many toxicologists, the resulting substances are not always any less harmful than the initial substances. Huge quantities of these sorts of deactivated antibiotics end up on fields through livestock urine (i.e., liquid manure). And once there, a considerable portion of them are converted back into the original active form. Researchers (Pollmer et al. 2001) fear that these antibiotics can be reabsorbed by plants through their roots because several months can transpire before the soil bacteria fully destroy the antibiotics.

No substance, least of all organic material, remains unaffected by this chemical catastrophe. The reuse of toxic or contaminated materials just perpetuates this catastrophe, and can even accumulate them in the worst-case scenario. Organic material from "outside," from the traditional, familiar environment, consists almost exclusively of plant and animal material that in all likelihood is made up of inferior proteins due to intensely chemically altered food. They can barely still muster resistance or sufficient vigor or regenerative healing power. If we offer our animals or our gardens and fields (whose soil life also consists primarily of animals, namely the edaphon, as we will see later) these "products" as nutrients and food—conventionally described as manure, fertilizer, or organic waste—what can we expect at harvest time? What is the use of trying to eat "organically" if we use organic material in our gardens or on our fields that was grown with artificial fertilizers, artificial foodstuffs, and pesticides? What is the use of the aforementioned scientifically conducted and government-overseen monitoring?

Half-Rotten Cropland

In *Wege zur Natur* (*Ways to Nature*), Raoul Francé (1924) describes a biological problem that already existed long before the negative consequences of our industrialized agriculture arose, and still hasn't been properly appreciated to this day—the state of mid-level rotting in our collective field and garden soil:

Just as sapropel [. . .] can be found in fresh water, rotten soil can also be found on land. [. . .] Mesosaprobic (i.e. half rotten, mid-level rotten) soil, on the other hand, is what is regularly found in fields and meadows in which the constant use of fertilizers continuously hinders its self-cleaning processes. Their functionality can only be restored if a period of fallowness allows the soil to resume its natural functions. This is in fact the most important function of fallow periods; they are a normal part of how the soil functions when its natural processes are disturbed by agricultural or gardening operations, so they have their place in garden soil too. Our fertilizers keep our fields, gardens, and meadows in a constant state of mid-level rot, no different than if one were to fill a pond with garbage and cadavers at regular intervals each year. This is not healthy, normal soil in the hygienic sense, and it's slowly getting to the point where we need to think about what kind of consequences this will have for our livelihoods. (34–35)

According to Francé, forest soil, just like any natural soil, has large quantities of "waste" available each year to be consumed via the curious phenomenon in which saprophytes, which feed on rot, take on a cleansing role that allows biologically fully purified, aerated soil that is free of rot to form. He continues:

"Pure soil" is characterized by a rich edaphon with the proper composition, with "aerators" and "deaerators" predominating. [. . .] But humans have encroached greatly upon nature [. . .] and made the soil arable, drained bogs and swamps, cleared forests, and then artificially maintained the resulting land in a state of half-rot by adding polysaprobic, often rotten material. What is generally known as "fertilizer" [. . .] is in fact a complete biocenosis of rot [. . .], a full community of microorganisms that feed on rotting material. Manure is a primitively prepared culture of bacteria, fungi, and flagellates—in other words, a true "edaphon culture" that is regularly applied to the soil. Garden compost is fundamentally the same thing.

Francé proposes "differentiating these special groups of edaphon and studying the processes of soil rot, soil purification, ventilation, and removal of rot, which are of massive significance to agriculture and hygiene, in the greatest possible detail."

These discoveries have also been ignored for decades. However, this would be an enormously important field of research.

SOMETHING HAS TO CHANGE

Abstaining from sealing the soil with concrete and asphalt is not enough to ensure its protection and fertility (Laukötter 2007). A much more destructive effect on our soil is caused by conventional agriculture! But this has hardly been acknowledged thus far.

Despite great agricultural challenges, despite food supply issues across the world, despite ever-worsening damage to the soil and groundwater, nothing is really changing because all new developments are reliant on an obsolete model: even today, all of the theories and practical guides in the field of agriculture are still built upon Justus von Liebig's 170-year-old *Organic Chemistry in its Applications to Agriculture and Physiology* (1840). Both academic teachings on agriculture and agriculture in practice, which is shaped by those teachings, still follow the false presumption that all living things can spontaneously form from their basic chemical elements.

Over the last 150 years, we have worn down and exhausted our soil—humus's habitat—through our agricultural practices that are based on this obsolete concept. Our agricultural methods are not only consuming what's left of our available healthy soil but also systematically poisoning and destroying it.

First we chemically poisoned the entire biosphere, and now we're in the process of turning it completely upside down through completely uncontrolled and uncontrollable genetic experimentation. Although the chemical compounds we've released into the environment (DDT, PCB, and so on) are inorganic substances, they have nevertheless found their way into everything, from the blood of Arctic animals such as polar bears to human breast milk, which is supposed to be the best nourishment our newborn children can get. And manipulated proteins in the form of DNA in living organisms can also multiply and carry out their own horizontal gene transfer, which means that the genetic material is not exchanged and passed on from generation to generation within a species (like it is through procreation) but moves "laterally" between organisms (including single-celled organisms) and even from species to species.

> **For They Know Not What They Do**
> Continuously releasing artificially modified organisms and genetic structures into the ecosystem and the entire biosphere (as we have been doing for quite some time now) without really being able to gauge the consequences, is as reckless as it is dangerous.

The main reason our field soil is becoming poorer and is ultimately losing all of its structure and natural functions (such as protecting against erosion, regulating water, and providing vitality and fertility) is the steady decline in and weakening of the life in the soil due to the exclusive employment of mechanical and chemical procedures in combination with excessive amounts of incorrectly stored (and therefore toxic) manure. This weakened soil yields weak harvests for weak animals and weak people. Molecules and chemical elements being quantitatively present does not on its own say anything about the life-sustaining nutritional value of a given food or about the quality of the soil that the food is grown in. The presence of living structures is at least as important.

Considering the problems that conventional agriculture has created—which we cannot (and must not) close our eyes to any longer and which culminate in the extensive and irreparable loss of the fertile humus layer in countless areas around the world—it is high time that we reconsidered things. We must no longer work against our planet's living systems—we must finally work *with* them. Even the very existence of the human species is highly dependent on functional, healthy soil!

Our Agricultural Methods Are Obsolete

Our agricultural practices are hopelessly outdated, and they are progressively poisoning the biosphere and our foods. Plants are provided with nutrients the way that the common perception and its corresponding concepts and teachings dictate. It claims that only minute, water-soluble salt ions are capable of penetrating the nearly impenetrable plant cell walls. So the theory states that everything, both organic and inorganic materials, must be composted (i.e., it must rot, decay, be "mineralized" into these small salt ions). But the method through which living material arises again in the cell out of these tiny particles, which only carry chemically defined information on nonliving material, has never really been researched.

There are also exceptions that refute this theory. For example, large molecules of growth hormones, "pseudo-hormones," and other substances can pass unhindered through leaves, stems, and roots into a plant's body by simply being sprayed, inducing the plant do things such as continuously grow until it dies. Other exceptions include carnivorous plants. Furthermore, there is evidence that plants, just like animals, can absorb amino acids, the building blocks for proteins. A further example is given by the very large biotechnologically manipulated molecule constructions that have "completely unexpectedly" shown up across all species boundaries in practically every sort of cell, including plant cells. And it's very rare to find descriptions of endocytosis in plants in the academic literature. And it is precisely here where the buried foundation of all biological agriculture, and with it the potential for nontoxic agricultural practices, in the future is found.

To put it simply, plants make use of all the preceding methods, which have been scientifically researched, in order to absorb material from their environments. Nutrients, synthetic chemicals, toxins, and irrelevant material: plants take in what we supply them with via water flow or gassing or spraying, just as they take in what they choose from the environment themselves. But crops are dependent on what the farmer or gardener cultivating them, in accordance with the doctrines he or she has been taught, provides them. And these days, a large portion of that is NPK minerals. As long as the plants keep consuming the NPK fertilizers, they keep getting supplied with more—with no consideration given to where they come from. In a worst-case scenario, they can even contain meat and bone meal and prions!

Ecological agriculture seeks to work against all that; but ecological agriculture, sadly, is armed with insufficient resources. The fields of microbiology and molecular biology are made up partly of very recent knowledge and partly of knowledge that is around a hundred years old but still hasn't been looked at critically by chemical engineers or even, unfortunately, by representatives of the ecological movement in the agricultural industry.

But there is another way. Hans Peter Rusch (1960) and his son Volker Rusch (1999, 2002) have given us an example to follow for plant nutrition and thus for agriculture and even for medicine. Hans Peter Rusch's model has become the foundation of organic-biological agriculture and even of ecological agriculture altogether. It is a cyclical model of living material that renews itself on its own, cleanly and healthily, without constantly accumulating waste and toxins through "cancerous" growth, as is done in our modern chemical-technological agricultural system (Vester 1987; see also page 115). But his basic model has been repeatedly abandoned, forgotten, or distorted. It is addressed in detail starting on page 46.

If We Don't Change Course Now, When?

Modern agriculture's great challenge is to address the issue of plant nutrition without prejudice. This means detaching itself from the 170-year-old view represented by the mineral theory and considering the endocytosis model, even though this model is yet barely known and feels unfamiliar. Remember the elephant analogy from the introduction.

Acknowledging the model of the cycle of living material is a definite precondition if we are to establish a viable concept of alternative agriculture that provides a new approach to growing.

— CHAPTER TWO —

Suppressed Knowledge and Forgotten Models

ENDOCYTOSIS: HOW LIVING CELLS ABSORB NUTRIENTS

Now we will consider an explanation of what comprises the model of the cycle of living material.

"Endocytosis" is the term for the absorption (or incorporation) into a cell of molecules of different sizes (or of viruses, bacteria, and other single-celled organisms) by bending the cell wall inward and then pinching off the so-called transport vesicle to transport the nutrient caught inside within the cytoplasm, the basic material of a cell (see page 40).

Endocytosis has long been generally known to be an important element of cell nutrition in animal cells. But in the context of plant cells, I first found the subject mentioned in a published

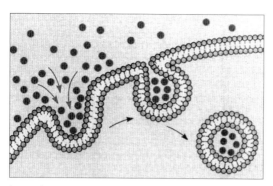

In endocytosis, the cell takes in molecules or particles by pulling its cell wall inward to surround them and then pinching off the resulting vesicle and pushing it into the cell's interior. The cell can use the same process in reverse to expel particles outward (exocytosis).

format in an interview with John Hamaker in Tompkins and Bird's *Secrets of the Soil* (1992, 187–198; German edition published by Omega-Verlag in 1998). *Secrets of the Soil* also talks about Bargyla Rateaver, who republished her work from 1973 in 1993 under the title *The Organic Method Primer Update*, in which she clearly and provocatively contrasted plant endocytosis with the mineral theory. Page 191 gives the following important information:

> An explanation for this startling cell absorption process was given by Mark S. Bretscher in December 1987 in the magazine *Scientific American*. In an article about animal cell locomotion, he describes what is generally known as endocytosis. It consists of cells bringing parts of all sides of their membranes inward. These small indentations take in the material in question, pinch themselves off, form vesicles, move into the cells, and, "IN PLANTS," there release their nutrients again, which are a thousand times larger than the ions in a chemical fertilizer solution—a difference on the scale of the difference between a mouse and an elephant.

The significance of this is not just that it provided new knowledge. What's truly new is that endocytosis, generally known from zoology, was also observed in plants. But even today, the endocytosis model for plant cell nutrition is stubbornly disregarded or contested by the bulk of the research community.

But plant cells—especially feeder roots, which are primarily involved in taking in nutrients, and their root hairs—are capable of absorbing anything as food via endocytosis, from the smallest NPK ions to large amino acid molecules to the largest protein building blocks of organic material. Plants are also able to take in viruses, bacteria, amoebas, and algae through the same process. This leaves only extremely specialized single-celled organisms (the so-called gas eaters) as pure primary producers, which build themselves exclusively from inorganic chemicals.

The endocytosis of living material by plants contradicts the idea that plant nutrition takes place solely through absorbing minerals.

This model supports the postulate (commonly found in older biological studies) that plant roots function like a reversed intestine, or, conversely, that our intestines are similar to a reversed plant root.

The more you look into plant physiology research from outside today's agricultural science, the more quickly you realize that the ways all living cells get their nutrients can be described with similar models, or sometimes even the very same model. Extremely similar or identical transport mechanisms for nutrient intake through cell walls can be observed in humans, animals, *and* plants. A simple comparison makes the presumptive evolutionary parallels clear, as the following figure illustrates.

The drawing portrays an example "original organism" that has formed from the amalgamation (endosymbiosis) of previously independent cells. It acquires its nutrients from various substances in its environment by absorbing them through the cell wall (corresponding in this model to the organism's surface) into its cytoplasm.

The "outer wall" or "outer skin" of a multicelled organism has been pushed in over the course of evolution. The gastrointestinal tract between the mouth and the anus forms along similar principles in the embryos of higher organisms. Here the organism must take in parts of the environment via eating in order to be able to transfer them through the cell wall or outer skin contained within the body (corresponding in this model to the intestinal wall) into its cytoplasm.

Comparison between models of plant roots and animal intestines

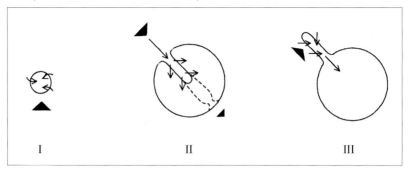

I. The drawing portrays an example "original organism" that has formed from the amalgamation (endosymbiosis) of previously independent cells. It acquires its nutrients from various substances in its environment by absorbing them through the cell wall (corresponding in this model to the organism's surface) into its cytoplasm.

II. The "outer wall" or "outer skin" of a multicelled organism has been pushed in over the course of evolution. The gastrointestinal tract between the mouth and the anus forms along similar principles in the embryos of higher organisms. Here the organism must take in parts of the environment via eating in order to be able to transfer them through the cell wall or outer skin contained within the body (corresponding in this model to the intestinal wall) into its cytoplasm.

III. It's equally possible, however, for the cell wall or outer skin to push outward and develop into a plant root. This outwardly pushed organ actively searches its direct environment for nutrients.

It's equally possible, however, for the cell wall or outer skin to push outward and develop into a plant root. This outwardly pushed organ actively searches its direct environment for nutrients.

The key aspect of these models is that they make it clear that plants can take in, digest, and add the same nutrients to their living bodies in the same way as other forms of life. According to current knowledge, all life-forms essentially depend on supplying their bodies with material that has already been prepared by other life-forms. This logically implies the model of the cycle of living material, which can be most easily visualized as eating and being eaten. Life-forms eat each other, and neither people nor animals nor plants are an exception to that rule. Broadly speaking, all forms of life require proteins, primarily in living form (i.e., proteins incor-

Do We Believe Certain Things Simply out of Habit?

The fact that plant cells are obviously so very comparable to animal and human cells in a physiological sense is something that unsettles the current predominant doctrine and throws it into question, but it must be the basis of future biological agricultural research. It opens up the way to a toxin-free system of plant nutrition and a chemical-free system of land and soil care, for which the pioneers of biological agriculture have been laying a foundation for more than 170 years.

porated into living systems). And these are very large molecules that, according to current knowledge, contain information vital to life.

THE REMUTATION MODEL

Recognizing that endocytosis takes place in plants is an important piece of support for the microbiological model of the cycle of living material, which includes microorganisms.

But there is also another area of microbiological research that seems

The principle of eating and being eaten permeates all forms of life—not just animals!

to have completely lost the attention of the modern scientific community. It essentially represents the second half of the endosymbiosis theory developed by Lynn Margulis and the adherents of the Gaia hypothesis (see page 65). This is remutation, postulated by Hugo Schanderl. By 1947, Schanderl had already succeeded in breeding and regenerating—remutating, as he called it—living, viable microorganisms out of certain cell components, such as mitochondria and chloroplasts, from plant tissue after it died. These experiments show that any living cell is capable of releasing new life after it has died.

Schanderl described remutation in agricultural soil bacteriology as follows in 1970: "When a plant is buried, the soil is enriched with bacteria not only because a vast number of existing soil bacteria decompose and break down the 'plant corpse,' multiplying tremendously in the process, but also because the soil is enriched with bacteria from higher plants as they break themselves down. Certainly, bacteria present in the soil also find abundant nutrients during composting, which allows them to multiply. But, as can be experimentally demonstrated, no bacteria need to enter from the outside whatsoever for decomposition to take place and a breeding ground of bacteria to arise" (Schanderl 1970).

He continues in the same article: "A significant proportion of the bacteria regenerated from plant cell organelles present in cow dung return to the planting soil. Unlike artificial fertilizer, this kind of fertilizer is filled with life and enriches the soil with bacterial life, increasing its fertility."

This would be a confirmation of my personal interpretation of the German word *verwesen* (to decompose), discussed in greater detail in the following box.

Life Arises from Life

Let's take a closer look at the German word "verwesen" (to decompose). What a remarkable word: "ver-wesen." A Wesen is literally a "living thing," while the prefix ver- implies a change in state or transition. If we consider this literal breakdown and carry it further, we might come up with the alternative term umwesen, with the um- prefix implying repetition, a new form, or a cycle. This would imply that a biological Wesens-Wechsel (change of state of a living thing) is taking place, instead of or alongside the chemical Todstoffwechsel (change of state of dead material), which is completely different from the wrongly named current theory of so-called mineralization.

After more than fifty years of being ignored and denied by the sciences, the remutation model is now being indirectly confirmed by cellular and molecular research (e.g., Kroymann 2010). Autonomous DNA (i.e., genetic material) that is independent of the cell's nucleus has been found both in mitochondria and in chloroplasts, which has led to acknowledgment of the endosymbiosis theory. In evolutionary terms, this also describes how ancient single-celled microorganisms relinquished their independence in favor of organizing into larger cells and, in a manner of speaking, were relegated into subordinate cell components.

Through the same process in reverse (in other words, through remutation), these cell components can also regain their independence! Schanderl's evidence, which he collected over the course of his whole life, was not accepted until 1970. I haven't been able to find any further publications about remutation since then either. Unfortunately, I never received an answer to a related question I asked Lynn Margulis (who died in 2011), although her writings mention an important source for her own research whom Schanderl was also familiar with, namely Ivan Emmanuel Wallin (1922, 1927.

Schanderl's remutation model implies that all decomposing organic substances, as well as all seeds that are starting the development of new life, are most likely capable of reshaping their own cell components into autonomous microorganisms such that living plants can employ their help—if they reabsorb them from their surroundings—to carry on their metabolic processes. The question also arises as to what extent living cells are even able to absorb an exclusive diet of inorganic, water-soluble salt ions (i.e., mineral fertilizers) without the help of microorganisms and, above all, process them into living material.

It should be clear that research on remutation and related areas (see Rusch 1968) is not easy because it is disputed, discarded, denied, or rejected ("filed away," as I like to call it). For example, Gunther Vogt (2000) attempted to do "file away" remutation in his collection of source materials (see also page 19), which was comprehensive but thoroughly one-sided in favor of the traditional interpretation. But what we should do is also reconsider and put to the test our conception of what a cell is.

Reconsidering the Cell Model

The traditional perception, which still prevails within biology, is that within an organism—a community of cells—an individual cell is the smallest coherent unit of life. From this it is inferred that when an organism dies and decays, its cells and their contents are surrendered to complete decomposition: they are "mineralized" into the smallest possible units. This is what all of us learn in school. But the research carried out by Rusch and Schanderl suggests something else. Both researchers provided large quantities of evidence showing that cellular substances and cell components such as the nucleus, mitochondria, and chloroplasts maintain and carry on their living characteristics in a sense after the organism's death, implying the existence of a cycle of this living material.

THE CYCLE OF LIVING MATERIAL MODEL

How should we envision the cycle of endocytosis and remutation? The following drawing gives a schematic depiction of a single plant root hair cell. At the top right you can see an endocytosis and exocytosis cycle in the living cell. The bottom left shows the remutation process researched by Hugo Schanderl after the end of cell organization which, as we have learned, is how we should think of the death of an organism. If we now bring these two observations together, we end up with the following depiction as an explanatory model for the cycle of living material.

In endocytosis, the cell wall bends inward and then forms a bubble-like "container" that carries the substances and particles located outside the cell wall nearby into the cell's interior. By using this method, plant cells take in nutrients in the same way as all other types of cells that are involved in nutrition in different forms of life. They take in protein molecules of all sizes and even bacteria and other single-celled organisms. This is the basis of my perhaps provocative claim that all plants eat meat. (I mean this ironically, of course, since in fact they actually eat vegetables.) As they travel back

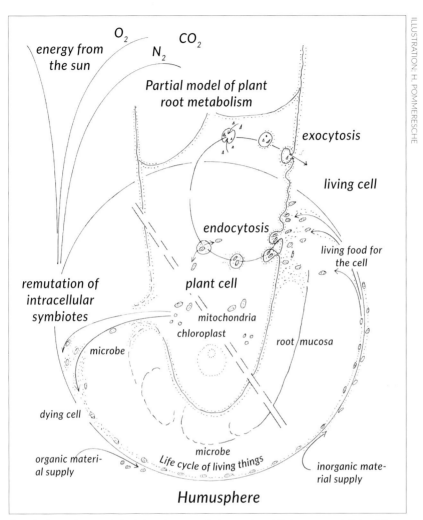

Depiction of a model of the theory of the cycle of living material. Is this theory unrealistic? Or just not yet sufficiently researched?

to the interior side of the cell wall, the vesicles also carry along residual material that is then released outward (through the reverse of the process by which they were brought in). This also causes the cell wall material that was used to form the vesicle to be returned back to the inner side of the cell wall.

In every cell that plays a role in nutrition, hundreds of these vesicles are active at any given time. The principles of endocytosis and exocytosis in animal cells have long been accepted, but when it comes to plant cells, there is still only a small number of scientists who consider it realistic or are conducting research on it.

Schanderl and countless other researchers have shown that the organelles of the cytoplasm, such as the aforementioned mitochondria and chloroplasts, "reconstitute" themselves and become autonomous after the observed death of the overall organism. They thus concretely and observably close the cycle of living material in the humusphere as the smallest living units, complete with all the biological information that can be contained within them.

The reverse process, through which the so-called archaebacteria (primordial bacteria) organized themselves into interdependent single and multicelled organisms over the course of evolution, is the core of Lynn Margulis's endosymbiosis theory within the Gaia theory (see page 65). This conjecture, that formerly autonomous archaebacteria must have merged together, was confirmed when DNA that was distinct from the nucleus containing them was discovered in mitochondria and chloroplasts. This also gave new currency to Schanderl's postulate of the existence of remutation, or the renewed independence of cell components (organelles).

Taking Up Forgotten Lines of Research

I consider the likelihood of a life-sustaining cycle of this type to be so great that I see it as the basis of a new form of agriculture that is in harmony with nature. And this is my appeal to researchers and to society: this is, in the truest sense, a gigantic, unplowed field for agriculture, plant physiology, and nutrition research that is lying fallow but wants to be cultivated!

Hugo Schanderl's research can be found in the academic literature between 1946 and 1970. Along with other researchers and citing studies that date back to the year 1875 (Béchamp 1875), he describes in great detail how to breed living bacteria that can reproduce from organelles of the plant's cytoplasm (the mitochondria and chloroplasts). He also shows that after the traditionally defined death of an individual plant, it does indeed cease to exist as a functional, cooperating system of cells, but the individual cells are then apparently able to release their organelles as autonomous, freely living microorganisms. But because these observations were not reconcilable with the then-accepted concept of monomorphism (conservation of shape and form) among microorganisms throughout the entire course of their lives or with the idea that they are completely destroyed (mineralization) after death, nature once again had to take a back seat to inflexible scientific theories, and Schanderl's remutation model was filed away as unscientific.

This research also confirms the stance taken by Hans Peter Rusch, who never wanted to believe that the most valuable and complex substances that nature can produce (like proteins) were utterly destroyed, decomposed, and mineralized after an organism's death.

And if we take this line of thinking a step further: if this cell material (the cytoplasm and everything else that it contains) can be passed on in living form from one being to another, so can everything else. That includes all dead material, unimportant material, important material, material of uncertain importance—and all toxic material. That all of this is mineralized, disappearing without leaving behind any trace and thus no longer being able to cause any harm, is something that I cannot believe.

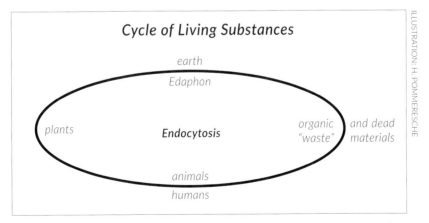

Schanderl's remutation model, with its continuation of living characteristics after the death of an individual, provides a comprehensible explanation for the long-suspected cycle of living material.

A NEW UNDERSTANDING OF PLANT PHYSIOLOGY

"Nature is whole and complete, it is only humans that pick it apart with our divisive knowledge, categorizing it into good and bad, useful and harmful" (Fukuoka 1988, 112); "and science rages endlessly on" (Fukuoka 1987, 35)—all the way up to chaos theory and the butterfly effect, a point where even scientists are beginning to suspect that they once had a healthy, cohesive world as a partner.

In the standard doctrine on nutrition and ecology, plants are portrayed as primary producers of nutrients. This means that plants build themselves exclusively out of basic chemical elements with the help of solar energy, thereby providing people with grass, flowers, and potatoes. Animals and humans form another, distinct section of the living world: they live off of grass, flowers, potatoes, and each other. According to this theory, which has been around since Justus von Liebig, plants feed autotrophically (i.e., only on inorganic substances); animals and humans feed heterotrophically (i.e., only on material from the bodies of other organisms).

As a result, plant nutrition had traditionally been considered fundamentally distinct from animal and human nutrition. According to conventional wisdom, this also serves as a scientific justification for providing plants, especially our vegetables, with NPK fertilizers. This distinction forms the basis of all modern agricultural technology. The large majority of biologists, farmers, and gardeners still follow this model. Because the idea of the mineral theory is so firmly entrenched in the thoughts and perceptions of agricultural biologists and agricultural technicians, it hardly seems possible to change that and to embolden them to consider a new model. But you *can* prove that plants don't exclusively feed autotrophically even using conventional scientific methods. Traditional science, in fact, provides a multitude of evidence that plants also feed heterotrophically. But so far the connections have not been made, and there has been no corresponding change in agricultural practices whatsoever.

From 1993 to 1994, Bargyla Rateaver brought together experimental results from a wide variety of traditional specializations and credibly showed that all plants absorb and digest large protein molecules and living microorganisms through their root hair mucous membranes through the process of endocytosis (already well known and scientifically recognized among animals). Together with the remutation model proven by Schanderl—showing that root hair cells growing in a sterile environment release some of their cell components as they die, which can then regenerate into living bacteria—this produces a model of plant nutrition that can function without any salt ions.

Although Bargyla Rateaver and some other researchers had long hypothesized that living cells from plants or animals had to feed in similar or identical ways, it has been only in the last few decades that concrete proof of this has emerged. However, since this proof is often found in the context of other research work, these fundamental discoveries have never reached a high degree of recognition. The established practices have also hindered the spread of this knowledge. It's even possible that certain lobbyist groups have consciously suppressed them due to their economic interests.

The Australian farmer Alex Podolinsky contributed a further building block to our knowledge of plant nutrition. He noted that plants have two separate root systems available to them: one for absorbing water, which is primarily used in transpiration and is thus a part of the plant's respiratory cycle (plants breathe, just as animal organisms do, and transpire water in the process); the second, the fine feeder roots, for absorbing nutrients. Podolinsky further observed that in plants fed with artificial mineral fertilizer, these "nutrient roots" are weakened or destroyed. He refers to a sentence from a lecture by Rudolf Steiner: "Plants should absorb elements through the humus, not through the water in the soil" (Podolinsky and Hatch 2000; Podolinsky 2008).

This model of two distinct root systems thus gives us yet another argument for why artificially salinized soil water weakens not just the life in the soil but also the plants themselves.

THE HUMUSPHERE: THE HABITAT OF PLANTS' ROOTS

The term "soil humus content" refers to the totality of all the organic substances present in the soil (see also pages 218–219 of the glossary for details). It is often expressed in terms of carbon content percentage, as carbon is the basic building block of organic material. But this definition is insufficient as it only reveals the sum of all the carbon atoms contained in the soil. How much of that is valuable compost, living soil biota, liquid manure, or other organic substances is not clarified. A relevant quote comes from M. M. Kononova's treatise, "The Soil's Humic Substances—Results and Problems in Humus Research" (1958): "The history of humus research is rich in incorrect approaches to clarifying important questions, which has led to contradictions and confused ideas about the nature of humic substances, their origins, and the role they play in forming the soil and determining its fertility."

But if we primarily understand "humus" as referring to the abundance of organic substances present in the soil, we overlook its mineral content. The proportion of minerals has increased in cultivated soils during our era in comparison with past eras in which consistently humid heat promoted the formation of organic soil material over huge swaths of forest for thousands of years. *The ratio between the organic and mineral portions of the material has shifted, to the detriment of the soil.* Incidentally, this is a particularly strong example of the importance of using the right terms with the right meanings: the word "mineral" is here used correctly to refer to everything from rocks, gravel, and sand to the very finest mechanically ground particles—it has absolutely nothing to do with NPK fertilizers or other salt ions (see also pages 224–226 of the glossary).

One small calculation is sufficient to get an idea of the significance of plant roots in the soil: "The formation of root hairs greatly increases the root's surface area. Rye (*Secale cereale*) has about 13,000,000 roots with a surface area of 235 square meters, and 14,000,000,000 root hairs with a surface area of 400 square meters in 1/22 of a cubic meter of soil. [. . .] The surface area of the underground portions is thus 130 times as large as that of the above-ground portions" (Jurzitza 1987, 28; see also page 110 of the color images).

One single rye plant has the equivalent surface area of *an entire garden* in direct contact with the soil in which it grows. What this soil is made up of has to be crucially important.

Annie Francé-Harrar (1957) wrote the following about how healthy soil should look for plant roots to be able to optimally carry out their work: "Ideal soil should have the following composition: 65 percent organic material, 20 percent edaphic organisms, 15 percent mineral substances. [. . .] But this kind of abundance of organic material exists hardly anywhere on the planet any more, the highest concentrations being in untrodden corners of tropical jungles, but never in our growing soil. But it is possible to restore the organic-inorganic balance in growing soil within a practical timespan through systematically employed humus management." These recommended ratios also provide a target to work toward in systematic humus management. But she was already well aware of how difficult it is to put this into practice: "But this [. . .] means a radical agricultural revolution, much larger than the one triggered by Liebig in his time" (20).

How Does the Humusphere Form?

Topsoil formation is very much a classic case study in the movement of living material from the waste material of living things into plants, of the descent of living material into Mother Earth. It's also a study in the soil, of its many functions, of its conversions and storage until its reappearance in the world of above-ground organisms. The bulk of the soil material first becomes clearly visible as nutrient-forming chlorophyll, but that chlorophyll would never exist without the work of the countless organisms in the soil.

The conceptual model of mineralization—the complete breakdown of all organic material into inorganic base materials—is first of all (and I cannot emphasize this enough) a technically incorrect use of the terminology. Second, it is logically improbable that it takes place, because that would leave only one possible explanation for the new life that forms, that being the concept of spontaneous generation, which has been rejected by the same scientific establishment.

The same species of bacterial symbionts appear in almost all animal and plant organisms, the lactic acid bacteria (see also pages 77, 165, 171, and 223). In fact, soil probes from all over the world, even if the soil in question is only slightly fertile, always contain large quantities of lactic acid bacteria. The soil contains more of them, and of better varieties, the more fertile it is. This is further evidence that the cycle of living material takes place in the topsoil through the mediation of bacteria. The remnants of biological processes on the surface, processed by countless species of small creatures, are first processed into precursors by budding fungi species, predominantly yeasts and molds, and then passed along to the bacterial symbionts in the soil. According to the most recent research, these symbionts—lactic acid bacteria in this case—can be directly consumed and digested as food via endocytosis by plant root hairs (Rateaver and Rateaver 1993), and they leave all kinds of organic material behind after they die, especially in the fall.

These particles, as well as the bacteria themselves (i.e., the living material in the soil bacteria), are a prerequisite for the formation of high-quality soil: topsoil that is aerated, loose, water-retaining, capable of biological tillage (Sekera 2012), safe from erosion, and fertile, the result of the functions of the edaphon, as outlined by Henning (2011). The adhesiveness of the microorganism residues cements the inorganic mineral substances of rock erosion into soil crumbs. In contrast to the views of agrochemists, it is this alone that deserves the name "humus" in the biological sense: a conglomeration of organic and inorganic material. And this means that it is completely impossible to describe humus as a dead, chemical substance!

Humus formation is a sort of "organic predigestion" for plants; and at the same time, humic soil serves as a pantry of living nutrients during the growing season, when plants can grow only if supplied with sufficient warmth, water, and sunlight.

Otherwise, however, the parallels between animal and plant digestion are unmistakable. In both cases, microorganisms serve as an intermediate station, as "nutrient facilitators," and in both cases organic or inorganic material can be extracted as needed from the nutrient substrate and used to build cells and tissues.

In purely spatial terms, the humusphere is the sphere between the atmosphere (the gas sphere) and the lithosphere (the rock sphere) and constitutes the biosphere together with the hydrosphere (the water sphere). In the humusphere, the entire metabolism of all dead and living material is carried out in a continuous cycle. It is driven by the oldest life-forms that we know of: microorganisms.

According to the cycle of living material model, we can attest that humus is created *by* life, *out of* life, *for* life.

THE EDAPHON: THE "RESIDENTS" OF THE HUMUSPHERE

Just as the water has plankton, there is also "the plankton of the soil," as Raoul H. Francé so singularly described and illustrated the term "edaphon" for us in 1911. The whole fertility of the humusphere depends on this edaphon, and it contains the entire existential basis of our life on this planet. The biosphere carries out its own cycle via its own living beings.

Under the heading "Ein Buch mit vielen Siegeln" ("A Book with Many Seals") in "*Mensch und Umwelt*," J. Filser (1997) writes: "Microorganisms in the soil can contribute to the nutrition of a plant or strengthen its resistance to pathogenic organisms or the effects of environmental damage. [. . .] They represent a biological potential that promises a wide variety of useful applications in protecting our resources in agriculture and forestry. [. . .] Thus far, only a limited portion of the soil microflora has been cultivated and examined in more detail. We are not even aware of an estimated 90 to 99 percent of soil organisms."

If we go back a hundred years to the year 1911, we can join Francé in a fabulous world of the very small that would astound us. He was the first to extensively describe the life in the soil, directly related to the topsoil and to humus. And he described it with the joy of discovery, using vivid language that can still enthrall the modern reader.

As already mentioned, ideas from Schanderl, Rusch, and Francé show up in Margulis's work from 1993 through 2001, albeit prosaically described:

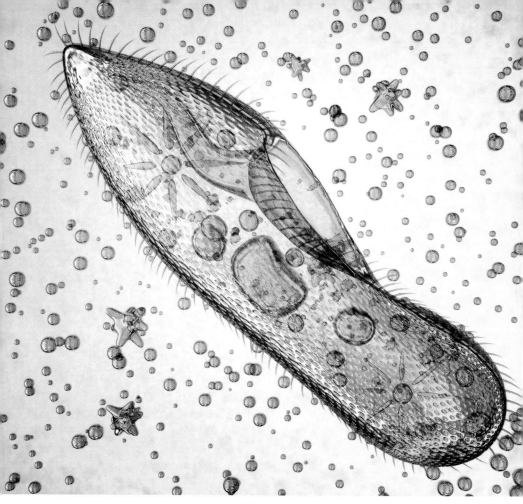

Paramecium caudatum, a single-celled organism that lives in fresh water.

Euglena spec. is a single-celled organism containing chlorophyll and a member of the "plankton of the soil."

> There is still little information to be found about the role of the split algae [editor's note: he is presumably referring to the so-called blue-green algae, now known as cyanobacteria] in agriculture. The fact that they play a very crucial role in the preparations for weathering was itself new to science [. . .]. You can perform the experiment yourself by hitting a seemingly completely unweathered gray rock face with an alpenstock. Soon a damp green spot will appear as a sign that there are plants living there. All rock and stone exposed to the air and to the rain is coated with a veritable lawn of split algae. It is known [. . .] that sandstone hides around 24,000 of these "weathering plants" on each square centimeter of its surface. It is impossible for that not to have a considerable impact. And it has actually emerged that these tiny plantlets [. . .] use special materials to turn the rock into a solution which they can feed on, although only if the preconditions for weathering are present, rain and sunshine. They are the first humus formers, and it is only once they have done their preliminary work that the lichen and mosses can take root, which are generally what is credited with causing weathering.

This is followed by a fantastical description of the enchanted forest, full of living gems, that one must read in the original to free himself from the NPK model and be able to think beyond it (Francé 1911).

Annie Francé-Harrar spent forty years working with Raoul Francé and published the book *Humus, Bodenleben und Fruchtbarkeit* (*Humus, Soil Life, and Fertility*) in 1957. She describes how the edaphon works to form topsoil even more extensively and in greater detail in *Leben wird aus dem Stein* (*Life Arises from Rock*; 1964, 2014):

> The final dissolution of the soil minerals does not, as was previously assumed, take place through purely chemical means, but rather in combination with the decomposing effects of microbes. The latter is the more important component of the process. Humus formation from the soil minerals is dependent on it. The Earth quite literally

has microbes to thank for being anything but pure rock. Incidentally, this also casts a new light on geologic periods. For a long time, before it could turn over the land to large animals and plants, the earth was having its living spaces prepared via the accumulation of fertile soil. Life in the form of single-celled organisms that are invisible to the naked eye long precedes multi-celled life.

This is the establishment of a common theme that would carry on from Schanderl through Rusch all the way up to Margulis and the Gaia hypothesis and endosymbiotic theory in the year 2001:

> All the [. . .] microbes stick in place in, on, under, and between the bulging, green to olive green pillow of *Nostoc* or the constantly gently oscillating filaments of *Oscillatoria tenuis* [editor's note: a species of cyanobacteria] in the Nile-green forest. [. . .] From the very beginning, zoogloea of one or multiple varieties form. In a certain sense, they are the common workplaces where the lives of all lithobionts play out. They protect against heat and cold and against drying out or being washed away. Surely a microclimate exists under this colloidal cover that is independent of the outside world. Temperature extremes on an unvegetated rock wall fluctuate between 60 and -60 degrees Celsius. But an algae cell is never frozen by such lower temperatures; they never cease their work.

A tiny, self-regulating greenhouse—an important part of the Gaia theory (see page 65).

As you might imagine, a world is opening up here that we have been aware of for over a hundred years, but it is a world we still know very little about and certainly not one that we look to in agriculture.

The Edaphon Is the Plankton of the Humusphere

Just as plankton provides the basic conditions for all further life in the hydrosphere (i.e., in the water) the edaphon provides the basic conditions for all further life in and on the soil. (Strangely, the term "edaphon" has long not been used as frequently for the life in the soil as the term "plankton" [from the Greek word "planktós," which means "drifting"] has for the things roaming around in the water.) Just as plankton—which you can take striking video footage of in clear water—is nothing more than tremendous quantities of living cell material, the edaphon plays the role of living cytoplasm in the topsoil. Of course, it is hardly possible to film it or to make it visible any other way, which may be one of the reasons why it is so absent from the general consciousness. But it's possible that even larger quantities of proteins are hidden within it than in plankton! A comparison accentuates this point: the average human biomass in the United States is 18 kilograms per hectare. On the other hand, the average biomass over the same area of insects, earthworms, single-celled organisms, algae, bacteria, and fungi is about 6,500 kilograms. That's more than 350 times as much! To break it down more specifically, that's 150 kilograms of single-celled organisms, 1,000 kilograms of earthworms, 1,000 kilograms of insects, 1,700 kilograms of bacteria, and 2,500 kilograms of fungi (Gaia 1985, 150; see also page 110 of the color images).

And what's the breakdown in our gardens? In a 1,000-square-meter garden, 1,000 kilograms of edaphon lives and works in complete silence, without disturbing the neighbors, all year long.

The biomass of the earthworms alone—who represent a part of the edaphon—can be determined by any interested layman by collecting and weighing them. In sandy soil beneath conventional barley, I've found 9 grams per square meter of surface area (to the depth of a spade), 49 grams beneath conventional pasture, and 840 grams in my own biologically cultivated garden soil. Per hectare, or 10,000 square meters, that comes out to 90, 490, and up to 8,400 kilograms of living biomass respectively. And that's just the earthworms!

The ideal for fertile growing soil is (Francé 1911, 1995): 1 kilogram of living biomass or edaphon per square meter of garden or field soil. One kilogram of living cytoplasm per square meter corresponds to 1 metric ton of biomass per 1,000 square meters, or 10 metric tons per hectare. Ten hectares of cultivated field soil thus come with 100 metric tons of living biomass in the form of the edaphon beneath the ground—and that's without even considering what can be achieved above the ground. That comes out to the weight of one thousand hogs or two hundred cows!

For readers who are farmers: these numbers reveal how much "livestock" you have hidden on your farm, without ever seeing or noticing it. If you converted this quantity into actual livestock, how much would you have to supply them with on your farm in terms of feed, stalling, ventilation, weather protection, and eventual excrement disposal? How about veterinary costs, consultation, and researching?

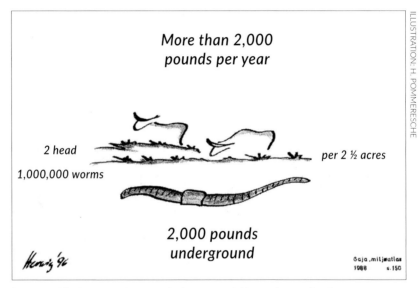

"Livestock" above and beneath the ground (from the author's Norwegian lecture materials).

This is only hard for us to conceptualize because we have neglected it in the models we use to teach and think about agriculture. If you think about these numbers in the context of the new, still unfamiliar "plants can digest protein" model, you quickly recognize the new possibilities that they reveal for our agricultural practices.

The literature and practical experience can give different numbers from those presented here, however. This has a simple explanation: in (field) soil that has been intensively chemically and mechanically treated, the living conditions for the soil organisms change massively (e.g., nutrients, habitat, soil structure), leading to the quantity of living edaphon dropping further and further. In soil that has been cultivated biologically, or in pristine virgin soil, on the other hand, the quantity can be significantly higher.

Anyone who wants to delve deeper into the still largely unknown "new community of life" that is the edaphon is advised to read the writings of Francé and Francé-Harrar from between 1911 and 1957. The two researchers take us along on an adventurous journey of discovery into the "underworld" of humus and of living things. Their fascination with and enthusiasm in their observations is downright thrilling. It wasn't until almost a hundred years later that something similar appeared, in the form of Lynn Margulis's *Garden of Microbial Delights* (1993).

The Edaphon Is the Best Possible Source of Nutrients for Plants

It is the edaphon that forms the soil structure and maintains it, through processes called biological tillage. The natural living processes of even these smallest of life-forms are in a continuous biological cycle. High-quality, erosion-proof soil with up to (optimally) 50 percent pore space is simultaneously both the consequence of and the indispensable precondition for soil consistency. Remember Francé-Harrar's definition of the ideal soil composition, possible in untouched nature: *ideal soil should contain 65 percent organic material, 20 percent edaphic organisms (the life-forms constituting the edaphon), and 15 percent mineral substances.* This requirement seems almost impossible to meet, as most cultivated soil on Earth today is far from it: modern field soil is composed on average of 1 percent edaphon, 3 percent organic material, and 96 percent sand and rock (of which the organic component is continuously being bombarded with more and more artificial salts and acid residues and decimated by pesticides). It has about 10 percent pore space (various studies call for up to 50 percent).

As farmers, gardeners, and garden lovers, we are treating the biomass in the soil the wrong way. We're continuously destroying the soil's structure, which is built up by the edaphon together with the plant roots themselves, with heavy machinery. We destroy any plant that we think we "won't need" with veritable chemical weapons, with the consequence that these plants no longer provide nutrients to the edaphon and can no longer fulfill their role in photosynthesis. We use sprays to destroy "pests" without realizing that by doing so we are seriously decimating the edaphon, year in and year out. The surviving organisms absorb the poison, carry it further, and finally smuggle it into the cycle of living material in the form of living (but now weakened and sick) cytoplasm, with it finally ending up on our dinner plates.

Nevertheless, we feed 50 metric tons of stinking, highly toxic liquid manure and 1–2 metric tons of salts and acid residues to the soil life per hectare of field soil. This directly poisons some portions of the edaphon—earthworms, for example—and greatly constrains others in their "soil work." The result: our modern chemical- and technology-oriented agriculture with its "harvest successes." We will take a closer look at how great these successes actually are starting on page 97.

Accepting the possibility that plants feed on the living edaphon (i.e., that they also "eat meat") provides an explanation for phenomena such as these: the results of Siegfried Lange (Haye 1985, 1993) and in Lau (1990, 2017), with 2 metric tons of winter rye (see pages 146–148), as well as the results of my own year-long experiments with 18 kilograms of onions per square meter (see page 176), are products of unusually deep, loose soil structures made up of large quantities of undigested organic material, which allows for the growth of proper amounts of fresh edaphon as food for the plants.

These results are also staggering: the historical hydrocultural systems (which I discuss in detail starting on page 136) can be shown, as in Smil (1997), to have produced harvest yields that (in contrast to our modern harvests that benefit from advanced technology) "[could] feed 5 to 15 people [. . .] per hectare." Presumably they exhibited similarly high proportions of edaphon and of organic material to the values given above. This can also be credited to well-developed edaphon, as we can find unusually good conditions for ample plankton growth in almost all of the ancient hydrocultural systems; in these systems, plankton is directly available

The gold in the soil: earthworm excrement is more than just "waste"!

The earthworm is the best-known "resident" of the humusphere. Their numbers have dropped precipitously in the soil in many places.

A large number of earthworms is a sign of healthy soil and edaphon.

Siegfried Lange's winter rye. Left: The washed-away roots of a rye plant. Right: The same, cut up.

Returning the forest to the fields: Finely cut chaff after ten months of fermentation on the left and ten years on the right. The chaff is mixed with the garden soil at a 1:4 ratio. Then these kohlrabi plants grew in the substrate and were supplied with 20 percent aerated stinging nettle liquid.

to the plants as a living, vitality-sharing source of nutrients (see pages 137 and 144 of the color images).

In contrast to these numbers is the following assessment: "With today's most modern chemical-technological resources, the USA produces enough food for one person and Europe produces enough for barely two per hectare of field soil. Intensive, done-by-hand agriculture is just as effective today as it was during the Stone Age in providing consistent, controllable harvest yields" (Hitschfeld 1995). Siegfried Lange's harvest results (Lau 1990, 2017) also confirm this.

Probing the Causes of the World Hunger Problem

Can it be that the difference between 1 percent and 20 percent edaphon, as well as the difference between neglecting and caring for these masses of living cytoplasm (which plants can directly "digest" via endocytosis), is connected with the five- to fifteen-fold greater harvest yields in the latter case just discussed? Is the conventional agriculture we currently practice really about biological sustenance and care, or is it in fact about chemically and mechanically exterminating 10 metric tons of life per hectare of agriculturally "cultivated" land?

The question is which we want to choose: a system of agriculture based on Liebig's purely chemical model or a system of agriculture based on organic microbiology in accordance with Rusch. Maybe the two fields can and must be researched and applied co-equally, but in a completely new and collaborative manner?

THE GAIA HYPOTHESIS

I believe that the concept of remutation has a promising future ahead of it, as remutation is the inverse second half of the endo-symbiosis theory from the Gaia hypothesis of James Lovelock and Lynn Margulis, which, to summarize it briefly, states the following:

all forms of life manage and maintain their common living space, the biosphere, through self-regulation and symbiotic processes.

In the case of plants, that means that ancient bacteria, the archaebacteria, merged together for the sake of cooperation or teamwork. In the process of doing so, they changed and partially gave up their special characteristics that enabled them to live autonomously in favor of a common, overriding structure—namely the cell—in order to form new, multicellular organisms, culminating in the higher plants, which even developed the ability to provide and maintain our planet's atmosphere with an oxygen content of 21 percent!

This Gaia model is also compatible with the fact that, as mentioned earlier, certain cell organelles (mitochondria and chloroplasts) contain DNA (i.e., genetic material) that is completely different from the nucleus's genetic material. This phenomenon is found not only in plants but also in animal and human cells, where it has been introduced even to the general public during recent years under the name epigenetics. The idea that these components were once independent microorganisms and that, after "fulfilling their duty" as organelles in an organism's tissue, they remutate (i.e., transform into their original form), allowing them to carry on the cycle of life, is not as far-fetched as it may seem. Quite the opposite: the most recent discoveries in stem cell research seem to point precisely the direction in which Schanderl was heading.

If you stop looking at the humus layer as "dead soil," but rather as the result of a process that exactly counteracts the breakdown of living material, an entirely new outlook presents itself. This very complex process, involving true minerals working in conjunction with the edaphon and the organic material contained within it, doesn't result in a chemically analyzable substance or mixture of substances—it creates an entire environment between the atmosphere and the lithosphere, which we call the humusphere.

And since a constant and self-perpetuating buildup of living material—not just a breakdown, as is conventionally believed—takes place in the humusphere, it is an essential piece of the overall biosphere because the entire metabolism of all biomass takes place in it. According to Gerhardt Preuschen (1991), virgin soil contains

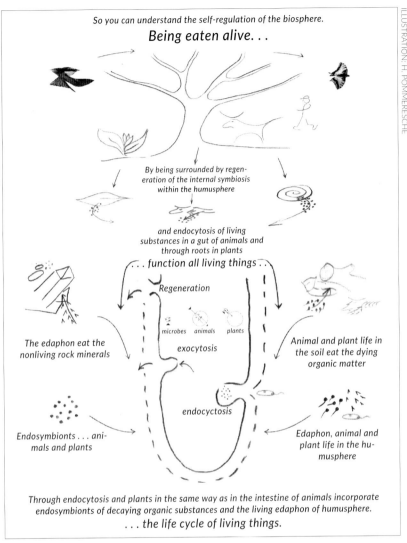

One way to visualize the biosphere's self-regulation.

practically no salt ions. That corresponds with the idea that most organic substances are passed along in the form of larger organic building blocks (such as proteins) or even as living nutrient sources in the form of bacteria, fungi, amoebas, and algae long before they are broken down into salt ions. They are not "decomposed" by the plants but rather "transformed" into new cell organizations, into new living beings, into flowers and vegetables, and then further into

> **Taking an Unbiased Look**
>
> The question is: Shouldn't we finally draw the right conclusions from our discoveries and treat our soil differently instead of assaulting it and ultimately destroying it with heavy machinery, salts, acid residues, and toxic, anaerobic liquid manure?
>
> My answer: Feed and care for your edaphon diligently and regularly with the purest possible food—just as you would feed and care for your farm animals above the ground. Try to maintain it and create the proper conditions for it to multiply, and it will build up and structure your garden soil to the benefit of your plants. The principles of microbiology will determine the future of successful agriculture and gardening.

animals and people, and after that returning into the humusphere, constantly completing new cycles.

In the next chapter, you will find out about the role microbiology has always played, and continues to play, in the soil (and in biology in general).

In the author's "rock garden," you can see how the edaphon is maintained over the winter: a cover layer of cut-up twigs (spruce, for example) protects the soil processes in the winter.

Feeding the edaphon. Right: Cut-up kitchen waste.

Right: Cover layer of cut-up twigs. Top and right: Good harvest in the author's onion plot without any dirt-digging work: up to 18 kilograms of onions per square meter.

Vegetables planted in living, humic soil, stimulated by effective microorganisms. Their edaphon is "fed" with organic material.

Half-finished compost is actually at the peak of its nutritional value.

This is what healthy soil looks like.

Experiments carried out by the author: the lettuce plant on the left is growing in "Åkerstedt mulch," the middle one in sauerkraut, and the one on the right in regular garden soil.

Chlorophyll water being produced in a blender.

— CHAPTER THREE —

The Significance of Microorganisms

MICROORGANISMS "MANAGE" THE HUMUSPHERE

The realization that plants actually release their own "living colleagues" and that these (together with countless other microorganisms in our biosphere) not only have played an essential role in the evolution of life on this planet but in all likelihood will continue to do so should serve as a wake-up call. We should take a closer look at the fellow inhabitants of our planet that we otherwise hardly notice but on whom our human existence depends more than we may even be able to imagine: the microorganisms.

Microorganisms are found everywhere in nature. Anywhere that fresh water accumulates: in every forest lake, in every stream, in every drainage ditch, in every temporarily flooded meadow (as long as it hasn't been kept dry by dikes), you can find countless microorganisms carrying out biological functions in the water and in the mud. And they are even present in the soil in the form of soil flora, or, to put it another way, the "population of the humusphere."

In habitats that are untouched by humans, humus forms in layers. For example, the humus formation carried out by certain groups of single-celled organisms might be unable to take place in the same layer as the work of certain other groups because the antigens (antibiotics) that are formed by both groups make it impossible for them to work together. The deeper humus layer, however, contains very few bacteria because the nutrients from waste material have already been consumed by then as soon as (or, better said, while) humus is formed. That the work carried out in these layers is dependent on the existence of a protective layer above them, the nutrient cover, is mentioned here only in passing. The fundamental consideration is that the fine plant roots, which perform digestion, never reach the soil's "work layers" (for a microscopic image of these fine roots, see page 110 of the color images). Rather, they reach just the microorganism-poor humus layer, in which they can—with the help of the photoautotrophic chloroplasts in their green leaves—synthesize carbohydrates that provide nutrients for the underground root flora that assist with their digestion. Through this process, the plant supports a system of root flora that is similar to the gut flora in animals and humans.

A soil layer is always unmistakably marked by the microorganisms that are able to live within it:
- putrefactive layers by putrefying bacteria
- fermenting layers by fermentation agents
- cellulose-breakdown layers by cellulose decomposers
- nitrogen-synthesis layers by nitrogen bacteria
- humus-formation layers and others by lactic acid producers

If we could only learn to stop constantly fighting these natural functions, often to the death, then we would once again be able to simply state what Alwin Seifert (1948, 1991), one of the pioneers of biological gardening and agriculture, asserted in his book

Gärtnern, Ackern—ohne Gift (Gardening, Farming—Without Poison): "Anyone who follows what is taught in schools about gardening, pomiculture, and agriculture today is (still) unsuspectingly wreaking havoc on his best helper, [. . .] the life in the soil."

WHAT OUR INVISIBLE FELLOW INHABITANTS DO

Rotting, Decaying, Fermenting: Anaerobic Processes

What fundamentally unites every aspect of agriculture is the transfer of material between plants and the environment. You can understand this biologically by examining three microbiological processes that play a crucial role in the transfer of material, if not causing it outright. These three processes are:
- anaerobic decomposition;
- aerobic decomposition; and
- fermentation.

Anaerobic decomposition stinks and is toxic. Aerobic decomposition smells like earth and the forest. Fermented material—such as sauerkraut—tastes good. At this point we should notice the contradiction that, on the one hand, stinking, toxic slurry is spread over fields, while on the other hand, good organic material and "spoiled" silage is thrown on the garbage pile.

Anaerobic decomposition (i.e., a process in which oxygen is not present) with temperatures that can exceed 60 degrees Celsius, warm compost, liquid manure, biogas in gigantic containers—almost every farmer produces these. There are specialized microorganisms that can convert material under these anaerobic conditions. They alert you to this special status of theirs through their stench. Between them, the products they create are toxic, combustible, or explosive. Their surroundings hide numerous pathogens and also attract them. We should ask ourselves: Why does society tolerate this stench? I also question why farmers choose to and are allowed to produce our foodstuffs in that stench.

Up to 70 percent of the nitrogen in stable manure is in the form of protein (Scheuer 1999, 20). Living cytoplasm is a delicacy for plant roots. Shouldn't we be allocating what little research funds are available to biologists instead of always just financing NPK studies? Even foul-smelling slurry—which has to be processed by *anaerobic* bacteria due to its lack of oxygen—is ultimately converted by *facultative anaerobic* microorganisms (meaning that they can live without oxygen but switch to "oxygen mode" under the proper environmental conditions) when the slurry is aerated such that it can be "eaten" by the edaphon, converted into further edaphon, and is finally able to be absorbed by plant roots. But the question remains of whether this toxic shock and the conversion must come at the expense of the plants' growth and health. Teruo Higa (2000) points to the biosphere-wide contamination with pathogenic microorganisms and pins his hopes on the symbiotic, life-regenerating effective microorganisms.

A further problem to which I have yet to hear any solution from an expert: What's the deal with biogas, gas energy, and "improved" manure like the kind used in biodynamic agriculture? The manure is missing its energy! Bioenergy from biomass, perhaps grown with artificial fertilizer—isn't that a contradiction? First sucking oil out of the North Sea, then producing artificial fertilizer from it, then growing biomass with it and distributing it to people's homes—in my view, this is a farce. All biomass—as long as it hasn't been contaminated with chemical additives by even the most minor industrial process—belongs back on the field soil; not in the form of substitute artificial fertilizer but as the large molecules created by nature that all heterotrophic organisms depend on.

Aerobic decomposition, liquid composting, compost piles, compost as surface mulch—all these are "calm" and unspectacular, rather slow-moving processes in which the substrate does not become too warm and doesn't smell bad, and through which less gas, energy, and material is lost compared to the aforementioned warm composting. In the first phase, these substrates, from a microbial perspective, are very active, so that they can initially have an inhibitory effect on the sprouting of the seeds and continue to inhibit growth for a while after that. When it's halfway decomposed, its effects are still active, and it is full of life and still full of nutrients. Later, when the compost is "finished" and the earthworms have left because all of their potential food has been eaten, then it's ready in the truest sense of the word: mild and friendly with everything sown and planted, which will soon require new food while growing. Mulch and surface compost, serving not only as protection but also as a source of nutrients for the life in the soil, must be replenished before the nutrients are exhausted and the edaphon eventually "starves."

In fermentation ("ferment" is an old-fashioned word for "enzyme"), proteins with a catalytic effect are active—meaning that they cause chemical reactions without being changed themselves. Fermentation is based on specific microbiological processes. Microorganisms generally metabolize sugar in an airtight container and primarily produce lactic acid and enzymes in the process. Very little energy and relatively few nutrients are consumed. The lactic acid fermentation of cabbage into sauerkraut is a typical example of this kind of process (grass silage for the storage of soilage and the torrefaction of grain feed are other examples). In Gerhardt Preuschen's (1991) *Ackerbaulehre nach ökologischen Gesetzen* (*Agricultural Doctrine According to Ecological Laws*), we find this principle described for storing stable manure with minimal losses. If a farmer is only going to come out to his fields for the purpose of feeding the edaphon once per year, he should do so with the best possible nutrient sources, as will be discussed in the next section.

STABLE MANURE AND SLURRY: VALUABLE PLANT FOOD OR POISON FOR THE SOIL?

Manure produced by any sort of creature smells disgusting and is unpleasant. But manure was the precursor of artificial fertilizers and is still used alongside them. The dawn of agriculture (in Europe) supposedly took place around ten thousand years ago. If we take off the chemical glasses that can only see nonliving, basic elements in everything and dissect anything more complex (even living) organisms, into inorganic basic material, we can discover true life, even in manure! So let's consider stable manure in a new light.

It's age-old wisdom that manure is necessary for anything to grow in your garden or your field from year to year. It's age-old wisdom that you apply manure to the soil so that the plants grow.

It's also age-old wisdom that manure stinks and contains and spreads disease agents.

Newer is the knowledge that untreated manure collected in large quantities emits poisonous gases (ammonia and decomposition gases) that can kill entire pig populations and even the farmers themselves and can blast the whole facility into the air if "improperly" ignited. Even newer still is the knowledge that raw, anaerobically stored, rotten (i.e., untreated) manure is also toxic to plants.

But the explanation is simple: it is since the advent of artificial fertilizers that we've come to see manure as a plague and a source of contamination. "That is smells" so unbearably like "spring" is caused solely by improper handling during storage.

There is a theory among farmers that manure contains the same components as artificial fertilizers, namely the things that chemists identify as vital to plants according to their theories and methods: nitrogen, phosphorus, and potassium. Chemists consider it unimportant where the NPK fertilizer comes from and whether it stinks, or even if it may be carcinogenic (Kristensen 1996). In any case, it's scientifically demonstrably NPK, right?

But if you stop looking at plants as purely chemically based, solely collecting and compounding together dead material from the soil? If you are aware of the large body of evidence showing that plants feed in exactly the same way as all of the other cells and collections of cells that we call organisms? If plants actually absorb living material? If plants absorb proteins and microorganisms through endocytosis, then they can also absorb and pass along artificially manipulated genes and infectious prions. Horizontal gene transfer exists, after all. Is there an alternative way to look at things?

The alternative is that the manure is meant as a source of nutrients for the life in the soil (the edaphon), which in turn serves as food for the plants. Slurry is frequently used even in organic agriculture—albeit preferably manure from one's own farm—but mineral fertilizers and sprays are not. But are the practitioners of organic agriculture really aware that in 10 hectares of land, they have 100 tons of living edaphon to provide for? How can they ensure that the edaphon flourishes and is able to multiply so that it can build up and maintain the soil's structure?

The cycle of material through the humusphere is powered by oxygen-loving (aerobic) microorganisms. Manure (i.e., an *anaerobic* organic substance stored in oxygen-poor conditions) needs to be processed by specialized microorganisms before its material can be integrated into the cycle and ultimately be made available to new life. In the natural humusphere, this is mostly done by organisms that breathe air (and thus oxygen). The cycle is maintained through ceaseless eating and being eaten. Most of this happens through the most direct route of being both eaten and passed on while alive, which is highly efficient.

If artificial salts are off of the table, is manure the best option that we can offer to our loyal helpers in the garden or the field? "Every follower of the current doctrine on gardening and agriculture is unknowingly causing havoc to his best helpers," Alwin Seifert wrote about the life in the soil as early as 1948, then again in 1970.

But we've continued to cause havoc. We constantly fall back into our old habit of fertilizing the soil, but we've never fully understood the feeding of the edaphon; and as a result perhaps we've never quite taken it seriously. But now the facts are plainly available: the edaphon and plants are full-fledged living creatures, and they eat the same things the same way as every other organism (Rateaver and Rateaver 1993).

The natural conclusion is that only the best is good enough for our plants! If we instead feed the edaphon with low-grade or artificial nutrient sources, it will not attain its natural vitality. Consequently, it will be unable to fulfill its duties: protecting against erosion, building up the soil structure, building up the humusphere with its exceedingly complex functions, maintaining the humusphere through the season, and directly serving the plants as living nutrient sources.

Instead of offering the plants and soil inhabitants toxic slurry and damaging chemicals, we should feed them with healthy, living cytoplasm.

THE PROBLEM WITH PRESERVATION

To give you a clearer idea of the importance of microorganisms throughout the entire biosphere, I would like to briefly address the issue of food production, yet another area where we find a strong connection to the life in the soil.

What Happens During Pasteurization?

Louis Pasteur regrettably taught us that we should fear all bacteria and microorganisms, a lesson that we continue to believe to this day and one that the food industry has found creative ways to sustain—very much out of necessity, as mass-produced food products piled on pallets have to wait days, weeks, or even years before they are sold. Pasteurized milk is filled with dead microorganisms. Most food products are boiled or treated with other methods, such as radiation, in order to kill germs, then made "safe" with additives and sold to us as packaged goods. This is unfortunately quite necessary in order to avoid a mass poisoning from the products being stored for too long. But all of these measures cause their own kind of poisoning, one that we don't notice until much later.

While food that has been pasteurized to death presents one problem, an additional one is posed by all the foods that contain preservatives. Do you really know what effect they have on your food? They are supposed to prevent food from spoiling. But what does it actually mean for food to spoil? There are fungi and other microorganisms that collect on food, just like on any other organic material, as soon as it dies in order to keep it moving through the cycle of living material. What we call spoiling is, from the perspective of nature, a completely normal form of conversion into new life. We call the things we cannot eat moldy or rotten; the things we can eat and taste good to us get names like sauerkraut, bread, and wine. And we find it again in our gardens in the form of compost.

All of these are the same conversion processes that microorganisms control everywhere. Preservatives should only kill the undesirable microorganisms that are harmful to our health. But what do you think actually happens to the living microorganisms in the fresh food and in our bodies? Preservatives are biocides (i.e., biological poisons). In the large majority of industrially produced goods made from organic material, all of the microorganisms are removed or killed with biocides. The fact that we still endorse these goods as food products is a serious problem. That we can digest it (i.e., microbially process it) at all is more of a miracle than a truly health-promoting form of nutrition.

And They're Preserved in the Soil Too

The resistance to digesting preserved foodstuffs that we find in our own digestive systems is naturally also present when organic waste is "digested" through composting. It decimates the edaphon and the biocides from our food remains enter into the materials cycles of our gardens and our fields, never to disappear again. The idea that they are broken down and thereby disappear entirely is not accurate. It is true that they can be broken down into their constituent parts. But these parts can chemically react with each other to form entirely new substances known as derivatives: substances that were not foreseen by anyone (compare also page 34 and the following pages). They ultimately show up "unexpectedly" somewhere, like dioxin (from strictly controlled, supposedly dioxin-free incineration plants) in breast milk.

Much worse are proteins from technologically manipulated genetic material. The material sits inside living cells and is promptly passed along through the cycle of living material, where, as part of living organisms, it can endlessly replicate itself. A single kernel of corn containing an industrially altered gene in your kitchen waste will maintain its functionality and have no problem passing through your compost to the microorganisms that break it down, which are then carried along to your chickens by earthworms, then finally into "organic" eggs and on to your breakfast table.

This process is known as horizontal gene transfer. Maximum safe limits unfortunately do nothing to help prevent it.

AN EXCURSION TO THE "HUMAN ECOSYSTEM"

It is impossible to overstate the importance of microorganisms in the ecosystem. As a demonstration of this, let's take a look at how microorganisms and humans coexist.

While it takes plants, animals, and humans weeks, months, or even years to react to environmental stimuli or changes in their surroundings, microorganisms show their effects after only a few days or even hours. This means that they can be used as indicators of health and biological quality in both medicine and agriculture. Or to put it a different way: microorganisms serve as the indicators of a living substance's biological quality, cellular fitness, nutritional value, and its overall health, and thus for the suitability of a given nutrient substrate or the quality of field soil.

Here we should again turn to Hans Peter Rusch: "The construction, reactivity, and specificity of living substances is so enormously varied and intricate that it is hardly possible to achieve precise scientific clarity through these means. Microorganisms can help us here. There are so many species of them, their natural order is so exact and exemplary, their reproductive ability is so great and yet at the same time their sensitivity is so high that this branch of biology, which has been so greatly neglected thus far [. . .] may be of great significance to both medicine and agriculture" (Rusch 1960, 60).

When it's intact, an organism's ability to fend off threats is a sort of "contract" between the organism and its living environment, represented by microorganisms, especially bacteria. The presence of symbiotic bacteria on our skin and mucous membranes is not only important for our metabolic processes, such as skin metabolism, but the bacteria also provide a certain degree of protection against the spread of pathogens due to the symbionts' ability to form antigens. But the most significant function of bacterial flora is that they present a constant challenge to the organism, one that it needs to prepare for and face, as the microorganisms can act as the most dangerous pathogens against organisms the moment they find a weak point in the organisms' defenses. This defensive system must therefore be vigilantly maintained, without any such weaknesses, since we will never be able to exterminate or fully fend off bacteria. The symbionts see to this, with their very presence (via their antigens) providing constant "defensive training." Since defensive capability, biological and functional tissue quality, and basic health are all one and the same, it is important to maintain a system of bacterial flora containing many of the symbiotic variety and none of the pathogenic variety. But that all depends on the nutrients available.

What Healthy Eating Really Means

"It's crucial for everyone, however, to provide the best and most valuable kind of nutrition that is available to the modern farmer according to the current state of science" (Rusch 1960, 63). Anyone who wants to stay healthy needs to continuously supply his digestive system with symbionts: fresh food whose cellular matter is fully functional because it originated in the soil, in animals, and in plants, and is healthy, vital, and not denatured by the absence of intact bacterial symbionts. Any processing, storage, and fresh preparation also need to follow these principles as closely as possible. The more living, vital nutrients we take in (or the closer they are to their original, living state), the more truly vital materials and life energy will be available to us.

Conversely, all nutrient sources grown with artificial mineral fertilizers and toxic pesticides have qualitatively negative effects and don't supply us with as much vital energy and nutrients as we need each day. Furthermore, along with our food we also take in the properties of the source plants and animals, which are weakened, poisoned, or genetically warped when subjected to technologically forced growing conditions. They pass along these undesirable characteristics through the entire cycle of living material, where they accumulate and strengthen their effects still further. With genetic modification, we face the danger of modified genes in living form spreading on their own through the biosphere. There is absolutely no way to control this sort of horizontal gene transfer!

And now we circle back around to field soil.

Organic Agriculture Needs to Be Our Food Source of the Future

Overcoming the difficulties of organic agriculture is only possible if you realize that it's a question of human existence, and that our existence is directly dependent on the cycle of living material.

EFFECTIVE MICROORGANISMS

Back to the very smallest of living things. There is a group of them that are known as effective microorganisms: particular strains that are cultured by humans. Are these miracle substances or a revolution? This "miracle" can be thoroughly explained via microbiology. Various types of effective microorganisms have long been available for purchase as delectable-smelling, sour-tasting cultures.

Since 1968 (also the year of the first edition of Rusch's *Bodenfruchtbarkeit* [*Soil Fertility*]), a group of scientists centered on Teruo Higa has been dedicated to effective microorganisms. Higa's book (2000) provides interesting reading to people who are familiar with Rusch. In it, Higa describes an illuminating incident: as a conventional NPK practitioner who didn't see any danger in the heavy use of chemicals, he became sick and weak, although he didn't draw a connection between the two for a long time. But one day he began to have doubts. He describes a major problem he encountered while growing watermelons:

> They were particularly badly afflicted with a viral illness that stubbornly resisted all of our efforts to get it under control. In the end, we had to concede defeat, and we removed the plants and threw them into the drainage ditches that the water from local kitchens flowed into. I forgot all about them until one day when I discovered that the plants were no longer displaying any signs of the disease that had plagued them and us whatsoever, and in fact had put down deep new roots, budded, and were already bearing fruit. [...] As I could clearly see the practical results, this became a convincing argument to me that I could not simply ignore. I think that this was the point when I finally became convinced that our agricultural system is too heavily based on the application of chemicals. I decided that I was going to look for a better way, possibly one in which microorganisms could be used to help with the plant growth. (Higa 2000)

Healthy plant growth is nearly impossible in compacted field soil.

Artist Carl W. Röhrig's depiction of the tunnels that earthworms live in.

So just as the reductionist scientist Justus von Liebig applied his chemical knowledge to agriculture on the one hand, we have in Higa a converted scientist who, like many of his predecessors, instead applied biology—the study of life—to agriculture instead. I think it's worth mentioning that he pursued similarly ambitious goals to Liebig in his time, and that just like him, he did whatever was necessary to reach them. Human history has shown time and time again that the vehemence and force with which an idea is publicized and spread has more to do with its success and acceptance than any actual consideration of what use it might have or damage it might cause.

While Higa promotes pure biological-microbial cultures, there are other producers that sell hybrid cultures of artificial fertilizer and microorganisms. For example, by 1995, TLB Syntrophic Microbial Fertilizer (TLB 1995) in China already had numerous active factories, with more under construction and quite a few others working under contract. Since then, genetic modifications have been described that work by "smuggling" foreign DNA into selected bacteria via infection with manipulated bacteriophages (viruses that attack bacteria). Furthermore, there are methods of deploying freshwater algae in rice paddies that greatly increase production levels. This is just a small piece of evidence that there are large-scale projects being carried out by the chemical industry that we here in Europe barely are aware of (or haven't been aware of for long).

But in the meantime, non–genetically modified effective microorganisms that were originally found free in nature have helped to start regenerating the soil in our badly strained world—and in the areas where it's happening, it has had great success (Higa 2000, 2004).

The Concept Isn't as New as It Seems

I think it's important to point out that essentially everyone who has been seriously involved in humus and compost research has put together special microbial cultures. For example, Annie Francé-Harrar described a way to seed with edaphon in 1957, and Hans Peter Rusch, a medical doctor who also wrote prescriptions for the humus, was very interested in the subjects of "symbiosis control" and "symbiotic therapy" in humans, animals, and plants. He developed a preparation called Symbioflor (Rusch 1968, 1991) that is still available today, in addition to other preparations that were also based on effective microorganisms as supplementary nutrient sources. (Author's note: research with effective microorganisms is also why we have reliable preparations for symbiosis control and symbiotic therapy in humans and animals in Europe; they reflect discoveries about the human microbiome around the world.)

But the international market is also full of microbial "fertilizers," which, to my knowledge, are supposed to be pure supplements and nutrients for the edaphon, whereas effective microorganism preparations are always proprietary, specially concocted mixtures of multiple types of microorganisms. Of course, they sometimes vary greatly in composition and quality from manufacturer to manufacturer.

What I would particularly like to point out is that we are surrounded by effective microorganisms essentially everywhere. Any living surface on a plant, animal, or human is going to be densely populated with symbiotic microorganisms. It's true of our skin (as you can read about in the previous section), of the surfaces of our eyes and teeth, and of our saliva and intestinal mucosa: all of them are thickly inhabited by symbiotic microorganisms. They are not only harmless but also absolutely essentially to our life, particu-

Microorganisms Are Essential to Our Life

We need to be clear about something: for every species that we wipe out because it is inconvenient to us, we are also wiping out a thousand others whose existence may have been dependent on that one undesirable species. Countless symbiotic, opportunistic, and neutral microorganisms are standing in the way of any number of pathogenic germs that could pose a danger to us—meaning that in most cases, we have absolutely no way of knowing what all we are actually exterminating.

larly to our immune system and its ability to defend against other, pathogenic (i.e., disease-causing) microorganisms.

Rusch's "cycle" and the practical instructions that stem from it are just as good a working hypothesis now as they were when they were first being researched. They have been completely and unjustly forgotten and suppressed. If you supply your plants with symbiotic microorganisms (see Higa 2000 and more recent authors), everything will grow great—inexplicably so to believers in the mineral theory!

Hans Peter Rusch specialized in the area of symbiotic microflora, and his son Volker Rusch and daughter-in-law Kerstin Rusch are doing advanced work with microbial therapy in the medical field. But the system of "microbial therapy for the soil" developed by Hans Peter Rusch as the basis of the entire ecological movement is no longer in vogue. But this principle applies to any effort to maintain health: as in the family, so in the stable and the field.

A crucial concept in plant nutrition and in our own (which you can read more about in Rusch's work) is that humans and animals digest their food in the stomach and the gastrointestinal system. The mucosae of the intestinal villi conceal the symbiotic intestinal bacteria that have a key influence on the entire metabolic process. Everything that we eat and digest is also eaten by these microorganisms. "Whenever we eat anything, we're passing it along for our bacteria to feast on" (Rusch 1955, 79).

But plants also have their own "intestinal villi," namely the root hairs on the fine branches of their feeder roots. Similarly to the intestinal mucosae, they surround themselves with root mucosae, which also host bacteria. It is these root mucosae and their bacteria that form the link between the part of the cycle of living material that takes place underground and the plants.

This Is where the Cycle of Living Material Closes

There is a piece of advice for our gardens and our fields that is often hinted at but almost never expressed outright and even less frequently explained convincingly: *in everything that you do in your garden, think in terms of life!* (This advice is even more applicable to the edaphon and plankton in our garden soil than to the higher plants, because the latter feed on the former.) As soon as we become aware of this new perspective, we suddenly start to notice other things too. For example, we realize that our rain-barrel water turns green after three days because it is full of green algae. Where do they come from? How do they get air to breathe? What do they eat, and how do they reproduce? How long can they survive in the barrel? Is it possible that they're also processing sunlight? And finally: Could they serve as a source of nutrients for my garden plants?

Microorganisms have been silently and diligently carrying out the aforementioned processes of building up and breaking down in the soil for millennia. Problems, but also possibilities, appear when people either pile on too high a concentration of waste or when they can no longer collect their food from the natural environment, instead turning to intensive cultivation, which will eventually involve mechanical and chemical measures that kill microorganisms. Similar reasoning also applies to the soil structure, cultivation depth, winter covers, and every other question that arises when growing things in a field or garden.

This is what people need to relearn!

— CHAPTER FOUR —

The Role of Water

WATER: CARRIER OF LIFE

In the search for living nutrient sources for the living, green part of our biosphere, there is one thing that everyone is aware of on some level but tends to forget: water is more than Coca-Cola or treated drinking water from the tap or from plastic bottles. Natural, living water has gone through a complete cycle: from the atmosphere as rain through rivers, seas, lakes, forests, fields, pastures, deserts, and mountains, across the humusphere to the lithosphere (or over concrete and asphalt into the drain) and out to the sea. Filtered by the humusphere and purified by the tireless "eating work" of the microorganisms, it wanders through rocks, sand, and clay and very gradually finds its way back to its sources, which distribute it once again to all living things via sunlight and the air.

Open water under open air is loaded with oxygen and nitrogen when it pours down into valleys through waterfalls. In addition to the sun's warmth, it also takes in particles and microorganisms from the air and riverbanks over the course of this journey. It turns into a "living soup" for microorganisms, which the humusphere desperately needs in order to carry out its work of stabilizing the soil structure, preventing erosion, purifying drinking water, and serving as the nutrient source for all green plants (without which we would have no oxygen to breathe). Thus, water is truly the carrier of life!

I've already mentioned several times the unusually high yields achieved by historic cultures, generally via hydroculture (see pages 140, 144 of the color images). This is another place where the chemical mineralization model does not provide a sufficient explanation. Understanding how these harvest yields, which are 5–15 times higher than those using conventional methods, are possible naturally requires some unusual conditions relative to today's standards.

These large harvests can be found practically in the Stone Age, among nearly all flood cultures and all other ancient users of hydroculture, as well as among the original jungle cultures. Dissolved salt ions are not found in appreciable quantities in the soil or the groundwater either among them or in modern virgin soil (i.e., soil that has not yet been touched or encumbered by human influence or anthropogenic handling). The few salt ions that are present naturally—such as those from urine—are quickly processed by microorganisms, which biologically bind them such that they (since they are water soluble) cannot be washed away by rain or groundwater. For the same reasons, the mineralization postulated by agrochemists—the breakdown of all organic material into its basic components or at least into water-soluble, inorganic compounds—does not take place, or takes place only to a very limited degree. Under such conditions, no proteins or organic structures would ever have been able to form and settle themselves on this watery planet.

Not All Water Is Created Equal

What is the significance of the differences between different types of water for plants and ultimately for agriculture? Francé may be the only one so far who has thought about and comprehensibly described water and the soil as a biocenosis, a biological community of all the life in the biosphere. I would also like to refer to Viktor Schauberger in this context. With his help, we can explore a wider world that many don't know about; for example, in the book *Living Energies* (Coats 1999), as well as in Schauberger's own works (2006, 2012, 2016).

Rainwater

In the past, rainwater was seen as the purest form of water that could be found. But the air, clouds, and rain have always carried large amounts of desert and volcano dust, pollen, and microorganisms (i.e., protein in living structures) as well as every imaginable gas around with them. Since even the smallest drop of the condensation that clouds are formed out of requires a cloud condensation nucleus that it "picks" from the impurities in the air, you can imagine the sorts of things that rain down on the ground these days. Among many other things, this includes several kilograms of nitrogen per hectare (i.e., artificial fertilizer). Also among them, however, is a considerable amount of living cytoplasm, which we have good reason to look at as the "atmosphere's sky plankton." I have been unable to find out how much of the atmosphere this constitutes.

Industrial Wastewater

By "industrial wastewater," I am not referring to process water, which is used directly in product manufacturing, but to drinking water that no longer merits that name. It is mechanically and chemically conditioned wastewater, which is delivered from the waterworks to households in the form of "completely harmless" and "drinkable" water. It is supposedly free of microorganisms, but as a result it is rich in chlorine, a biocide.

Groundwater

Groundwater can be the purest water that can be found, if we're talking about naturally pure water. Nowadays, however, it is so heavily polluted with contaminants in many areas that it needs to be handled with extreme caution. In any case, groundwater no longer reaches us in its original form from practically anywhere; instead, it is pumped to the surface and transported through pipes.

Spring Water and Soil Water

Spring water is groundwater in its original form. As impure rain and surface water (see the next section), it goes through a very particular closed cycle through the humusphere and the lithosphere, after which it reaches the surface again in the form of high-quality spring water. This water, which has gone through a complete natural cycle, does not contain any salt ions from nature: "The content values given by the chemical soil analyses that are common today have no relation to the soil's fertility, but they do provide important information about disturbances in the soil. Soil water must be biologically pure, without soluble minerals. If minerals are detected, the water purification process has been disturbed. The more minerals you can detect, the greater the stress is on all the life in the soil." (Preuschen 1991, 70). Here, the word "mineral" is being used with the same common, traditional definition that is used everywhere else, which I criticized earlier, leaving open and ambiguous the question of what is actually being dissolved in the water.

This brings me to another thought: water that is transported out of deep wells that were bored using technical means has only traveled through half of its underground path. The ascension, which it accomplishes by itself in the case of spring water, is accomplished mechanically here. Is it not possible that this bypasses important purification and energizing processes? (Compare the publications of Gerald H. Pollack [2015] and Viktor Schauberger [2006, 2012, 2016].)

Surface Water

Surface water comes from underground sources after a long natural cycle through the lithosphere, or it collects from rainwater in mountains, seas, forests, and bogs, gradually joining open water flows and ultimately rivers and streams.

Glacial Water

This water, which comes from (volcanic) mountains, exhibits an unusual characteristic: it contains true minerals in the form of rock dust or "stone meal"—the glacial milk, ground off of the rock surface and bed by the heavy glacier. The term "glacial milk" refers to the milk-like murkiness of this water. But it may have received this name early in the history of agriculture due to the high fertility that the glacial milk provided to the soil in the valleys where it settled.

Fundamentals of Hydroculture

In agriculture and gardening, many experiments are carried out with the similar, often mechanically produced stone meal. This ideally natural stone meal can serve as the source of life for, among other things, the lithobionts (i.e., the "stone eaters" among the microorganisms, described by Annie Francé-Harrar and Lynn Margulis). These rock inhabitants are probably among the very oldest bacteria. Francé-Harrar writes very explicitly: "The final dissolution of the soil minerals does not, as was previously assumed, take place through purely chemical means, but rather in combination with the decomposing effects of microbes. The latter is the more important component of the process. Humus formation from the soil minerals is dependent on it. The earth's crust quite literally has microbes to thank for becoming more than a rock." (Francé-Harrar 1957, 46)

With this description of these processes, Francé-Harrar comes quite close to the description that Margulis formulated in her endosymbiosis theory. This rock dust, as true, ultrafine mineral particles (and not as "dissolved" mineral components), is what the endocytosis theory claims can be directly absorbed and digested by both microorganism and plant root hairs (Rateaver and Rateaver 1993). And beyond this stage, very, very little is further broken down in the sense postulated by the mineral theory. The overwhelming majority is passed on while living or as close to living as possible through metabolic processes as the largest molecules possible in accordance with the principle of "eating and being eaten." This is not only the process that allows new life to continuously emerge but it is also the process that powers the entire revitalization of the biosphere.

— CHAPTER FIVE —

Putting it to the Test

How Economical is Agriculture?

AN HONEST EFFICIENCY ANALYSIS

You often hear the argument that says, "Protecting the environment and toxin-free agriculture; sure, that's all well and good—but it's much too expensive." In this section, let's take a look at the economic efficiency of the various methods. You will be surprised by the results. My reflections go beyond the idea that we have ripped the life out of the soil with machine energy and chemistry. I also ask how economical our system really is if we take into account all of the energy-intensive factors that have to come into play before the harvest.

Energy Expenditure for Food Cultivation: From the Stone Age to Today

The authors and figures given here are the primary sources of my knowledge. If needed, they take things substantially further. There are also countless other sources on the same subject. Overriding and determining everything else, however, is the desire to verifiably differentiate technological and biological agriculture, and to examine attitudes toward the turning point: further onward in an ever-faster journey into the abyss or coming to a stop and consciously working in accord with nature. I hope the following description contributes to our understanding of where these two possible routes have led us so far. At the same time, I want to show how important it can be to engage with our existing knowledge about both organic and inorganic substances and how necessary it is to conduct further research on them.

In the past, our ancestors applied their intuitive wisdom to agriculture. This is true of virtually every advanced culture (see pages 137, 140, and 144), which without exception were built on an adequate and secure supply of agricultural foodstuffs. We cannot investigate every culture here, as there are too many. I have selected representative examples for which we have relatively reliable information. For example, Vaclav Smil assembled five-hundred-year-old comparative data on Egypt, China, Mexico, Europe, and North America. The numbers show that Egypt, China, and Mexico all conducted highly specialized agriculture, with resultant yields that we have never even been able to approach with any of our new "accomplishments." It has been discovered that these cultures grew enough food for four to five people on each hectare of agricultural land. In the twentieth century, the United States and Europe grew only enough for one person over the same area. This hasn't changed much in the United States, while Europe was able to feed only one to two people per hectare in the twentieth century (Smil 1997, 117). You can calculate things differently here, but the results always tell us the same thing:

Since the introduction of technological and chemical resources to agriculture in the middle of the twentieth century, we have calculated and calculated and not only deluded ourselves, but we've also promised ourselves and the entire world a green revolution that has indeed increased yields by 50 percent, but at the cost of a 450 percent increase in outside energy. But the "forgotten" advanced cultures had 400–500 percent higher yields over the same areas! They were also many times more energy efficient, as we will see.

The Ratio between Energy Expenditure and Harvest Yields

Let's take a purely numbers-based look at what we've accomplished in the area of energy usage since we transitioned from a gathering culture to a culture reliant on chemical agriculture. The calculations are given as examples and, when not otherwise noted, were performed by me.

1. Prehistoric times

As hunters and gatherers in ancient times, people expended 1 unit of energy in terms of their working power (i.e., of their life energy) to catch or collect 10 units of food. *The in-out ratio was thus 1:10.*

Incidentally, a similar ratio has been noted among animals. It was observed that "a pair of chickadees with two broods of 10 to 13 bird offspring over the course of the summer collects up to 75 kilograms [. . .] of insects, eggs, and larvae from trees and branches" (Blotzheim 1993). Animals are still content with that today. Of course, whether they find enough "pests" for their food is another question.

2. The beginning of sedentariness and agriculture

When human beings turned their hand to agriculture about ten thousand years ago, that was an actual green revolution! From three-field crop rotation, rice paddies, and extensive irrigation systems all the way up to the floating gardens of the Aztecs (see page 140), yields exceeded 50 units and approached 100 while maintaining the same expenditure of human energy.

3. Twentieth-century, gardening families

I used ten years of documentation from a family of European gardeners to calculate a ratio of expended to harvested energy of

1:122 (an average annual input of 40,082 kilocalories and output of 4,877,820 kilocalories; Feist 1954). But more on that later (see page 150).

4. Twentieth century, industrial food production

If we use the same method to calculate in energy units as things developed in the twentieth century, we get the following data:
- In 1910, 1 unit of work energy was required to produce 1 unit of food energy in the United States.
- In 1945, the same amount of work energy produced 0.5 unit of harvest.
- Since 1970, only 10 percent of the input and the modern deep-sea fisheries and cattle fattening generate just 5 percent of 1 food energy unit from 1 energy unit.

If you extrapolate these energy calculations further, a disastrous progression becomes clear: as progress increases in the industrialization of agriculture, the in-out ratio of energy input to energy yield breaks down catastrophically.

As mentioned, deep-sea fisheries apply 1 unit in to receive 0.05 units out; the same ratio applies to hamburger production. For tomato production in greenhouses, it has been calculated that the energy value of 120 tomatoes is necessary to produce one tomato. Here, the in-out ratio is just 1:0.008. Or to put it another way:

Input = 1 unit of energy.
Yield = 0.2 or 0.05 or 0.008 units of energy.

From 122 harvested energy units per unit expended, we've thus arrived at a yield of 0.008. Economists call this negative growth. In global energy terms, however, it's a downright insane waste of energy. The question thus becomes: Do we save energy, or do we use it to destroy life? "The machines moved to the fields, and the farmers to the factories," as Bill Mollison, one of the founders of the global permaculture movement, explained (Mollison and Holmgren 1984).

To calculate the whole thing from the other direction: if we assign the mass production of beef cattle the value of one, we get an energy efficiency that was up to two thousand times higher than today's among the Aztecs (see page 139), and that a family of gardeners today, using purely human energy, generates up to twenty-five hundred times as much as industrial agriculture with all of its high-performance vehicles. Are these figures not at least as surprising as the figure of 10 metric tons of living biomass per hectare of field soil (see page 59)?

In every description of the development of agriculture from the last fifty to sixty years (i.e., as long as massive quantities of chemical and technological "aids" have been used), even in those written by ecologists, the sharp increase in harvest yields is interpreted as proof that technology-based practices are necessary. But if you consult the sources and scientific calculations that form the basis of these claims, you will accordingly find that the average increase in yields over these fifty years is 50 percent, which corresponds to a factor of 1.5. In the same studies, however, you'll also find evidence that the expenditure of energy has *increased* by a factor of 4.5.

Energy Balance: Ten Years with a Family of Six Gardeners\

The aforementioned family of six gardeners feeds themselves from a surface area of 2,000 square meters and kept precise records of it for ten years. Oswald Hitschfeld cites them very extensively in the out-of-print book *Garten, Heim, und Gewinn* by Ludwig Feist (1954). Obviously, the following numbers are example calculations with rounded and idealized figures and averages. But the key here is the orders of magnitude in comparison with our modern food growing using conventional methods—and they are impressive.

In an average year, this family produced almost 5 million kilocalories. To achieve this, the father worked an average of 32 minutes per day, the mother 23 minutes, and the eldest child 16 minutes in the garden. Those times add up to 71 minutes per day.

Twenty-four hundred kilocalories per day is enough to keep a person healthy and well. That comes out to 5.2 million kilocalories for six grown people in a year. To get through one 24-hour day, we can calculate like this: 2,400 / 24 = 100 kilocalories per hour.

Therefore, 71 minutes of work in the garden requires 118 kilocalories per day. This is the amount of energy one must expend in order to grow the bulk of his food in a garden, and he needs 333 square meters to do so. (Our technological agriculture "achieves" the same on 2,000 to 4,000 square meters.)

One liter of gasoline contains about 7,800 kilocalories of energy. One liter of gas is enough for a pleasant Sunday morning drive to the bakery to pick up bread, from a cold start. If you divide 7,800 kilocalories by the 118 kilocalories needed for a day's work in the garden, you come up with 66 days' worth of energy that you used for your drive to the bakery.

And because that was too many numbers at once, here is the whole thing again in a table.

Overview of Energy Expenditure

Per person per hour	100 kilocalories
Per person per day	2,400 kilocalories
71 minutes of garden work per day	118 kilocalories
1 liter of gasoline	7,800 kilocalories
1 year of self-sufficient garden work	43,000 kilocalories
6 trips to the bakery by car	46,800 kilocalories

This is an example from day-to-day life that can directly make clear what I described above: *people were once 2,000–3,000 times as energy efficient as they are today.* And this family of gardeners shows us that we can be too.

How Efficient Is Agriculture Today?

When the Ice Age came to an end in Europe about ten thousand years ago, soil formation—or better said, humus formation—took place via a succession of plant communities (tundra and so on) through the forest. However, the deforestation process taking place from the Middle Ages through today has largely interrupted the humus formation process. The consequence was a steady

Intensive potato production and mechanical rice production represent a "success" with their energy ratio of 1:5, meaning that one unit of energy expended mechanically (i.e., energy introduced to the system from elsewhere, or outside energy) produces five units of food energy. In intensive grain and soy production, the ratio shrinks to 1:2, meaning that one unit of energy yields two units of product. Milk production in the United States in 1910 was at just 1:1—barely an even energy balance. All the values above that mean the harvest or energy yield is less than the amount of energy expended.

This illustration gives a summary of what was said above, namely:

- That "our" gardener can be 2,500 times more energy efficient than deep-sea fisheries or the industrial cow fattening operations used by hamburger chains; and
- That you need a hectare of farmland in America today to produce enough food to supply one person; in Europe, just enough for two people is produced.

If we put together what I've explained so far, we can see that the people of earlier millennia were very efficient. And it's readily apparent that this efficiency is continuously declining: because the more machines and chemicals—in other words, the more energy, since it all must be manufactured and brought to the fields via the application of energy—we use in agriculture, the less food per unit of area we produce. The egregiously high amount of energy used today has brought machines and chemicals to the fields and the farmers to the factory halls (where they are now being replaced more and more by machines anyway).

This is the enormously negative growth that has been brought about by extremely elevated levels of machines and chemicals, including genetic manipulation, and that is presented to the people of the world as progress that will supposedly solve the problem of world hunger.

Yields Equal Quality?

So what do all these example calculations have to do with biology? The more energy (expended via mechanic methods, fertilizer production, biotechnology, etc.) we introduce into biological processes, the lower yields we end up with. And not just in quantitative terms but rather, as we will see later, qualitatively as well. Or to put it another way: the more technological energy required, the less biological yield.

What attentive, conventionally working farmer, or what agronomist, has not noted that more problems with "pests" appear the more agricultural toxins one uses? It's not just the old ones—new "pests" are appearing as well. Before the large-scale application of toxins in agriculture, mites were hardly a problem at all, and bacteria and viruses were the exception as well. But what accounts for their increase?

Chaboussou (1987) showed that agricultural toxins, even if they are pure contact poisons, always find their way into the plants and have an effect on their metabolisms. It's like sand in a gearbox. Even in small dosages, toxins can lead to disturbances to the delicate metabolisms of plants. For example, using herbicides can lead to the appearance of "pests" and using fungicides can trigger insect infestations or other diseases. Even an unbalanced application of mineral fertilizer can lead to stoppages in the plants' metabolisms. In modern agriculture, young plants receive too much fertilizer, and then later too little. In both cases, their metabolisms are disrupted because the "plant nutrients" provided via this fertilization method are absorbed by the plant along with its "drinking water." This leads to a destabilization of the plant—and probably to all other organisms as well, causing them to become sick. In my view, the plant diseases we've observed have their roots in classical agriculture's mistaken understanding of plant nutrition, because the plants are harmed by the chemicals and lose their sustainability (according to Chaboussou, due to disrupted protein synthesis). This keeps the plants from being able to adequately carry out their share of work with and in the biosphere.

The preceding concept also leads to the conclusion that our crops have forfeited their nutritional value to us as food—and that this loss has gone unnoticed, because it is not yet detectable with the current scientific methods. My thesis is if our crops and food plants still grow even though our field soils are severely poisoned, it is not because of these wrongheaded methods, but despite them! The humusphere with all its inhabitants is still only just surviving, which at the end of the day is pretty astonishing.

What we humans can achieve with our life energy is just one side of agriculture. The other side is what nature provides us. It can lend us its support and its help—if we are willing to accept this help. So far, however, we have simply consumed nature's resources without replenishing them (and in some cases we have already depleted them). What are we going to do when the resources that nature still holds ready for us finally run out?

Obviously, not everyone can or wants to set up a self-sufficient garden. But if those who can and want to do so, that represents a contribution toward solving the impending ecological and agrobiological problems—and it's also fun and fosters social cohesiveness in our society, something which we desperately need to cultivate. Numerous urban gardening projects, from communally maintained neighborhood gardens to small farming collectives, have appeared in recent years and are enjoying great popularity, confirming my view that my vision is not just a daydream anymore. In the appendix, you can find a selection of projects of this sort.

THE BIOSPHERE AS A MODEL: A PERFECT ECONOMIC SYSTEM

A good question to pose to children (or to be posed by children) is: Why isn't the forest full of leaves all the way up to the treetops, considering how many new leaves fall each year? Over the millennia, microorganisms have maintained a healthy biosphere according to the economic concept of nature, as Frederic Vester describes it in the following box—with zero growth in a steady state, without centralization and without monocultures, but with diversity in a decentralized association. The product line over the entire time period has consisted of a constantly developing collection of hundreds of thousands of different plants, animals, and microorganisms. The waste materials from this enormous production are pure breathable air and water vapor, the emission of mild warmth, birdsong, the rustling of leaves, and pure spring water.

Everyone knows the term "recycling" today—we stole the idea from nature. But what we're mostly doing is just recycling our own problems. After all, as we saw in the last section, our modern agricultural methods with mass production and high-performance machinery are not nearly as efficient as we tend to believe. But how can we produce food more efficiently again? Here we can also learn from nature, because it works more economically than we do, due largely to the help of the microorganisms in the biosphere, hydrosphere, and humusphere. To just give a hint of what the "power of the smallest organisms" has to do with us, let's turn our attention to Frederic Vester's *Wasser = Leben* (*Water = Life*; 1987), described in the following box.

This is Frederic Vester's "biosphere" super-factory, which—aside from humans—has issues with neither raw materials nor waste materials. Shouldn't we look to that as an example?

It is already practically proven that a person can grow all the food he needs in a year on only 330 square meters of garden (Hitschfeld 1995). He accomplishes this by introducing energy in the form of a small part of his human life energy. A reminder: the amount that he must employ for this is comparable to the amount contained in one liter of gasoline.

These facts come from science; it is all documented and proven. Nevertheless, we have created a social system in which these facts are ignored—and I view this social system as suicidal, because it not only disregards the principles of the cycle of life but destroys them as well. How long can that carry on?

"But is it even possible to cultivate the soil without applying NPK fertilizers and pesticides?" you ask.

It is possible: as long as you nurture and multiply the edaphon so that the plants have healthy, living cytoplasm to "eat" and can thus healthily grow their plant bodies themselves, without having to suffer under the constant, downright narcotic pressure of the

An Economic Overview of the Biosphere

Materials output: 200 billion metric tons of organic material, 100 billion metric tons of oxygen, several billion metric tons of light and heavy metals per year.

Raw materials consumption: None. Constant recycling keeps the mass constant at 2,000 billion metric tons of biomass.

Energy usage: 8,500 megawatt-hours of locally harvested solar energy per year. That's more than can be generated by a million large nuclear power plants.

Product selection: A constantly developing range of hundreds of thousands of different plants, animals, and microbes.

Pollution and Garbage: None.

Gaseous emissions: Breathable air and water vapor.

Radiation: Mild warmth.

Noise: Nature sounds.

Wastewater: Clear drinking water.

Zero growth due to being in a state of dynamic equilibrium. No central direction and no monostructures; decentrally organized diversity instead. Zero unemployment.

artificially added salt ions dissolved in their "drinking water." And as long as you try to learn from nature. The examples in the practical section (starting on page 159) will hopefully encourage you to try it yourself on a small scale.

Like earthworms, wood lice also play a role in the soil's ecological balance.

We rarely see it and only think of it as something dead, but the soil under our feet is teeming with life.

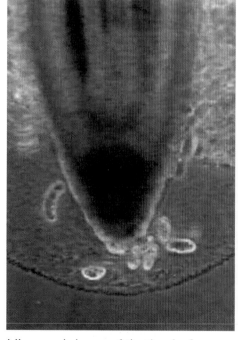

Microscopic image of the tip of a fine root hair.

— CHAPTER SIX —

The Agriculture of the Future

THE BEGINNING OF "ALTERNATIVE AGRICULTURE"

Plant nutrition (and with it all of agriculture) needs to be handled differently.

Let me summarize what we've learned: without the operators in the soil, without living material, nothing would happen with fertilizer salts—the soil would just break down. This means that if we exclusively supply our field soil with inorganic material, we will destroy it over time—provided it's even still in a healthy form as it is—without realizing it or intending to. The full cycle of living material reaches throughout the humusphere, in which the edaphon, the "plankton of the soil," forms the soil into its crumb structure, thus maintaining all of the humusphere's functions while simultaneously serving as food for the farmer's fields, pastures, and forests. The humusphere not only determines yield quantities but also the quality in the form of either a living, healthy harvest or a ruined, low-value harvest. The most important thing is, as the microbiologist and doctor Hans Peter Rusch emphatically urged, though it is not heeded in modern agriculture: *bring your in-house toxin production to an end*. But this can only be understood if you concern yourself with microbiological life processes.

The Most Important Publications

The cycle of living material model needs to be developed and publicized through every possible means. The most important works on the subject include:

- Hugo Schanderl (1947, 1964, 1970) on the remutation of cell components into living microorganisms that are capable of reproduction (he presents hundreds of experiments related to this theory);
- Hans Peter Rusch (1955, 1960, 1968–1991, 1969) on the preservation of the cycle of living material;
- Bargyla Rateaver (1993) on plant endocytosis;
- Lynn Margulis (1999) and the Gaia theory on endosymbiotic theory, "a new look at evolution";
- Volker Rusch (1999) on microbial therapy; and
- the works of Günther Enderlein (Krämer 2006).

And more and more researchers have been working on these topics in recent times (Kroymann 2010 being a very recent example).

According to everything that the researchers who are named here (researchers who have hardly been given an audience thus far) know, almost every form of life on this plant—plants, animals, fungi, and single-celled organisms—needs organic molecules, substances that have already been assembled or arranged by other organisms (Margulis and Sagan 1993). It is safe to assume that the cycle of material originating from living organisms (e.g., proteins, but also entire living microorganisms) is essential to the preservation of life.

All historical agricultural peoples (see pages 136–146) were dependent on artificial water supplies for their extensive agricultural practices. You can examine (and reconstruct) them in terms of the plankton and edaphon content of the water and soil substrates and compare the available quantities of living cytoplasm, of edaphon in the field soil, to the harvest yields they were able to achieve. At the same time, you can collect modern harvest records and investigate the available edaphon. In every case, you will find heightened quantities of proteins embedded in living substrate, which means directly available plant food (through endocytosis), a greatly increased pore volume, higher water capacity and erosion resistance, and not least of all greater yields per unit of area.

And then there's the matter of the biological qualities of the harvested food, which cannot be defined purely by chemical-analytical methods: qualities such as taste, disease resistance, shelf life, and above all good health benefits. Microbial agriculture can produce this sort of food, as has long been done in microbial therapy in the form of probiotics.

A conception of the living interrelationships in plant nutrition that is based on the microbiological model would substantially improve the state of biological agriculture in our society and would lead to a reformation of our entire doctrine on nutrition. Once again (as in the introductory chapter), I quote from Hans Peter Rusch (Rusch 1974) in his article "'Mineralisation' der lebenden Substanz" ("'Mineralization' of Living Material"): "Our understanding of metabolism has fundamentally shifted over the course of the decades. Anyone who still speaks of a total breakdown of all food substances into mineral salts has either slept through this entire eventful period or is pursuing certain mercantile goals that have nothing to do with true science."

How is it that this insight has still, in the twenty-first century, barely found a foothold in the field of agriculture? And this despite the fact that the research conducted by Rusch and Schanderl from the 1920s through the 1940s has been repeatedly confirmed in the meantime (e.g., Kroymann 2010; Paungfoo-Longhienne et al. 2010, 2013; Samaj et al. 2006).

As difficult as it is for this subject to find inroads into the scientific establishment—because it is competing with a very solidly entrenched and never scrutinized doctrine—it is quite easy to practically test it. All that is needed is to sufficiently and regularly "feed" the edaphon while refraining from chemical and mechanical encroachment into the humusphere to the greatest possible extent. Small-scale practical experiments and sources on historic and modern larger-scale projects are described in the last section of this book (starting on page 159).

Research in plant physiology also needs to fundamentally change its understanding of plant nutrition on the basis of the knowledge of endocytosis through plant roots (Rateaver and Rateaver 1993). This will inevitably lead to completely different practical methods of dealing with the collective system of life, which is made up of plants, soil, "fertilizer," nutrients, animals, and humans. This will allow for a description of the closed cycle of living material and of microorganisms in the humusphere. This has always been the endeavor of the truly life-logically working biologists, as I call them. Sadly, they just haven't found a receptive audience yet.

Farmers were once able to feed 5–15 people per hectare each year (Feist 1948). Today, in the United States—despite top-quality technology and genetic engineering—one hectare of cultivated land can only feed one person (Smil 1997). Records still survive that provide proof of harvest results from ages past (Hitschfeld 1984, 1995; Lange in Lau 1990, 2017). From the perspective of traditional agricultural doctrine, these harvest results seem to be beyond the realm of possibility. But anyone with an understanding of the principle of endocytosis in plant roots can readily comprehend these figures.

Just as plankton are broadly acknowledged as the foundation of all further life in the water, the edaphon, the plankton of the soil, serve as the foundation in the humusphere for all further life on the ground. Both working together under optimal living conditions is a sure sign of ideal plant nutrition through endocytosis and correspondingly favorable yields.

This is where I see the potential to finally establish a scientifically grounded system of agriculture that is truly based on biological principles. I also believe that all other research and technical fields (about which I am consciously not commenting on further) should be pursued and incorporated as enrichment of the microbial model. We don't need to completely discard our chemical-analytical model either. But it needs to be very critically reconsidered and scrutinized with regard to the concept that only life can create new life and that damaging or outright destroying living "participants" in the cycle of living material in the humusphere is harmful and contrary to the laws of biology.

What Are We Waiting For?

Evidence like the composition of virgin (and by definition ideal) forest soil or calculations of the contact surface area between root tissue and the soil—635 square meters in the case of a single rye plant (Jurzitza 1987)—should be proof enough of the kinds of yields that will be possible if the principle of the cycle of living material finds its way into agricultural doctrine.

"MICROBIAL THERAPY" IN GARDENS AND FIELDS

Chemical-technological agriculture has led to the poisoning of the entire biosphere, our food, and us: we find DDT in breast milk, and a variety of contaminants have been detected in the polar ice caps—far from any farm or settlement—while pesticides and their decomposition products are present in the deepest freshwater deposits, which seep into the groundwater over the course of decades, just to name a few examples. A system of agriculture grounded in microbiology, in contrast, would not only be able to regenerate our soil and our groundwater. It would also grant us enormous increases in harvest yields, giving it the potential to even solve the world hunger problem. When the cycle of living material is applied, it makes a radical "microbiological therapy" possible for our field soil and thus for our food. Even followers of ecological agriculture haven't factored in this aspect yet, because even among them the subject of plant endocytosis—the idea that plants should be recognized and understood as heterotrophic organisms—is not yet being discussed on a fundamental level.

Just to give one of many examples of "soil therapy:" potatoes that are infected with viruses heal completely in healthy, living soil; the viral disease is no longer detectable after a maximum of three years (Chaboussou 1985).

Though it may seem immaterial to a gardener or a farmer at first glance whether his plants feed on salt ions or on amoebas and algae, the impact and possibilities of these findings are of revolutionary significance to agricultural practices as well as to the soil that our food is grown in.

If you start with the premise that plants are heterotrophs, you inevitably come to the conclusion that exclusively supplying the soil with artificial mineral fertilizer is essentially a very particular form of synthetic "force-feeding." It has catastrophic consequences not just for the humusphere, but for the entire biosphere and thus for life itself as well. It should therefore be reduced and ultimately given up in favor of a truly biological system of agriculture.

Maybe our farmers are simply unable to make these necessary adjustments anymore. If so, we must turn to the gardeners—or else create an entirely new profession.

In addition to the new biology and plant physiology, examples like *Jedermann Selbstversorger* (*Self-Sufficient Everyman*; Migge 1918), *Der Gärtnerhof* (*The Garden Estate*; Schwarz 1974), and *Ökologisches Bauen* (*Ecological Growing*; Krusche 1982) have already developed ideas for alternative agriculture, for a truly organic form of agriculture, for the symbiotic agriculture of the future. You can find more on this in the last section of this book, starting on page 189.

Cutting to the Chase: What Should We Aim For?

The cycle of living material model gives us the biological understanding we need to properly care for the biosphere, and it is the cornerstone for the development of the agriculture of the future, as well as for a health-nurturing system of nutrition and lifestyle. With the help of the microbiological discoveries collected and presented in this book, we must strive to create a humus medium that (as a final goal) is composed approximately as follows (in proportions of the total weight):

- 20 percent edaphon (instead of 0.1 percent)
- 65 percent organic material (instead of 6 percent)
- 15 percent mineral substances (instead of 94 percent)
- 50 percent pore volume (instead of 10 percent)

Two minimum requirements must be met in agricultural practice to achieve these numbers:

1. No measures may be taken that inhibit, disrupt, interfere with, or damage the cycle of living and nonliving material. This means that organic fertilizer must be applied preferably without storage and without being forced in or even using deep plowing, because abstaining from this largely protects the layers of humus formation. Natural minerals, including trace elements, may only be supplied under the directive of the life in the soil, and all forced fertilization with nitrogen, superphosphate, and so on must be renounced.

2. No plant may be used as a source of nutrients or as feed if it would not be able to survive in that location without artificial toxic aids, or that would not be viable after breeding.

— CHAPTER SEVEN —

Organic Agriculture and Alternative Gardening

A CHANGE IN PERSPECTIVE ON "ORGANIC"

Why doesn't ecological agriculture (in its research, instruction, and practice) make use of the microbiological model of the cycle of living material? In my view, "organic" is not really "organic" unless the soil is cultivated on the basis of this model.

Truly Organic: What Does That Really Mean?

Nowadays, consumers are unable to really choose the food they buy, farmers have no proper instruction on how to produce it, and researchers cannot really determine how to produce ecological, biological food without the cycle of living material as a reference model. My view on this conundrum follows.

Taking a Critical Look

Farmers and scientists, as well as the laymen among us, have an incorrect idea of what ecological products and organically grown or manufactured food really are. It's not enough to look at ingredients like vitamins, enzymes, or antioxidants. These substances do sustain our bodies, but is that enough? The whole is greater than the sum of its parts!

"Organic" means "living." Nowadays we've forgotten what that really means. The fundamental mistake that was made in the history of (ecological) agriculture was that *the difference between dead and living material was disregarded*. Without knowing and accounting for this basic principle of the cycle of living material, a truly ecological or organic system of food production is impossible. Food that is truly organic is healthy and living, and it passes on its life energy—and not just the sum of its ingredients—through that vitality.

Organic Agriculture Is Humus Agriculture

For anyone who wishes to understand what organic agriculture really is, delving into the works of Hans Peter Rusch is indispensible. And if organic agriculture is to be the salvation of the worldwide hunger threat—which in my opinion it can be, and indeed must be—then the treatment of humus must be reconsidered, because organic agriculture is humus agriculture.

I found the following sentence in a book review: "Lynn Margulis (Gaia) has restored life to biology." There's no better way to express my impression that modern biology has become so completely dominated by biochemistry and mechanization that it is fundamentally no longer living up to its calling of describing life. The history of alternative agriculture, on the other hand, is the history of the cycle of living material. My hope is that it can also return ecological agriculture back to its actual organic roots. A substantial level of contact, a material connection between soil, plant, animal, and human is crucial. Humans and animals are directly or indirectly dependent on plants and the plants in turn on the soil, where the excrement and remains of all organisms return to, thus closing nature's cycle of living material and energy.

But living substances need to be understood as more than the mere sum of a collection of nonliving materials, molecules, and ions—even if this understanding is what you always hear and read. Living substances build themselves up, multiply, and decompose into new organizations of organisms throughout the humusphere in a cycle. There have been many attempts to create an explanatory model based on this cycle of living material using a variety of different approaches. *Der Kreislauf der Bakterien als Lebensprinzip (The Cycle of Bacteria as a Principle of Life)* by Hans Peter Rusch (1950) is one example: "The traditional doctrine on cells and tissues was based on the hypothesis that many chemical-physical barriers exist between nutrients and the organisms that ultimately consume them, barriers that could only be penetrated by easily soluble micromolecular substances. This hypothesis has never been directly proven."

No, it has simply come to be seen as fact over time. Rusch (1960) continues: "In fact, the entire doctrine on plant, animal, and human nutrition over the past decades has been brought into line with this concept. Our advances in amino acid, vitamin, and enzyme research haven't even been able to fundamentally change that."

The predominant scientific theory reduces nutrition into a simple question of addition: as long as an organism gets the proper mixture of amino acids, monosaccharides, and various fats, vitamins, enzymes, and trace elements, it will be fine.

Even now, agrochemistry has not progressed beyond the nutrient model centered on NPK and a few trace elements—for understandable economic reasons—and there seems to be no desire to move away from it. Where these so-called nutrients actually come from is, in principle, irrelevant. According to the theory, the only way to incorrectly feed an organism is by not providing it with essential nutrients, even if those nutrients are solely present within inorganic chemical compounds and can only be identified via analysis.

Third-Stage Substances

Hans Peter Rusch (1969) explained what is missing in this conventional model:

> "You can summarize the entire development of the study of metabolism so far in a simplified form as follows:
>
> **First stage:** nutrients and fuel (minerals, proteins, carbohydrates [sugar-like substances]), fats, large amounts of trace materials.
>
> **Second stage:** active substances (vitamins, ferments/enzymes, hormones, and a smaller number of trace elements).
>
> But in addition to these nutrients, fuels, and active substances, science has discovered a third large group, namely the
>
> **Third stage:** living material or the building blocks of living tissue. Any organism made up of living cells and tissues.

These third-stage substances have been known for a long time, but they were not incorporated into our understanding of the materials cycle under the assumption that all living organisms were indeed dependent on the nutrients, fuels, and active substances of the first two stages, but exclusively built their own living material out of these items themselves. This view has determined our entire collective understanding of metabolism, and every conventional model of plant, animal, and human nutrition is based on the concept. But it is in fact these third-stage living substances that determine what happens with the dead material. They ensure that its caloric energy is put to use, make muscular activity happen, cause the conversion of solar energy in green plants, and so on. They are in charge of practically every process that we think of as part of life. Living materials are essential catalysts without which there could be neither metabolism nor life. They are the "inhabitants" of the substances that we know from chemistry. They are the operators and the managers, the true kings of our biosphere.

The most important aspect of these third-stage substances is that they must be passed along through the metabolic cycle in living form in order to fulfill their life-sustaining and life-directing tasks. This is the source of the cycle of living material model.

Or to put it a different way: every living organism requires matter and energy in the sense that we learned in school. But in all likelihood, they also require a third thing, something that can be neither seen nor measured using our current methods: an organizing, governing force that allows the matter and energy to work together properly. We must presume that this force, this "governing authority" is found within living material, and according to the current science, it can only be passed from organism to organism or from organism to nutrient source in living form. Too little of it leads to weakness and degeneration. If none is present at all, the result is death. Life can only arise from life! A system of growing and producing food that is based on this knowledge and on a corresponding relationship with the humusphere is the true and original form of organic agriculture.

Natural foods always contain large organic molecules from the cells, tissues, and fluids of the food's source, predominantly organelles and microsomes from the cytoplasm, by-products of the breakdown of blood cells, and chlorophyll from green plants. But these macromolecules or living materials are not simply biologically undifferentiated, uniform substances but rather complexes whose formations and biological functions are dependent on their origins—life passing on its living functions. If they originate from a healthy, biologically intact cell, they will be equipped for the purpose of functioning as nutrients, which has to be done intracellularly. But they cannot fulfill this function if they originate from cells that are not intact or are sick (e.g., spoiled, in the truest sense of the word, by pesticides). Rusch explains this idea well:

> Where the nutrients come from is absolutely not irrelevant. If the nutrient substances originate from intact cells, tissues, and organisms, they will lead to very healthy cells and be capable of physiologically nourishing cells, tissues, and organisms, and even to increase the biological value of cells whose material is no longer in an optimal state due to sickness. But if they come from a sick or weak organism that has only had salt ions available and has had to be "protected" its whole life long, the biological value will inevitably decrease, and the constitution of the cells, tissues, and thus the organism will weaken. (Rusch 1960, 51)

In modern plant and animal nutrient production, no one questions whether these organisms are even still able to carry out their natural functions of self-preservation and reproduction—a determiner of their biological value—without the help of artificial aids. Rusch continues:

> The hypothesis that only micromolecules that are soluble in water, weak acids, or lye are ever able to make it through the "bottlenecks" in the metabolic cycle, that cell membranes represent completely insurmountable obstacles to large organic molecules, has been weakened by subsequent developments in the direction of more accurate nutritional doctrine. [. . .] But if macromolecules of living material do in fact take part in the cycle of nutrients, this implies that any living tissue cells formed remain intact after the natural death of the tissue, ready to participate in the cycle of living material as an extracellular organic entity and thereby be brought to newly-forming cells and newly-forming tissue and organisms. [. . .] This is the scientific basis for viewing all of the organisms that interact with humans in our environment collectively, as the source of the direction of human life processes. [. . .] If functionally intact, biologically active nutrient material is transferred from organism to organism rather than just lifeless, micromolecular material with no consideration of its origin, then, in the long term, all the organisms on Earth are completely dependent on each other to stay

healthy. So anyone who uses biologically low-quality nutrient material, regardless of whether it comes from plants, animals, or humans, is inevitably also decreasing his own biological quality because cells lose more and more of their biological functionality whenever they incorporate low-grade components. (Rusch 1960, 53, 56)

Nutrient substances, including their living components, rarely move directly from organism to organism. More often they pass through countless varieties of microorganisms along the way, which we might call nutrient intermediaries (microorganisms in the intestinal mucosae of humans and animals and the root mucosae of plants).

Future Research Approaches

Let me once again explain the key distinction between the mineral model and the cycle of living material model.

Liebig's mineral theory operates on the assumption that any material must be fully mineralized before plants are able to absorb it as food. This is called autotrophy, meaning feeding exclusively on inorganic material.

Rusch's theory, on the other hand, views this idea as unlikely. He found it biologically inconceivable "[. . .] that creatures would expend the effort of building up such complex substances only for them to collapse to functionless dust once the creature's physical existence comes to an end" (Rusch 1955, 106). Sooner or later, the inevitable conclusion of this line of thinking had to be that plants are also heterotrophic, meaning that they are dependent on metabolic products or living components of the bodies of other organisms.

If you want to clearly distinguish between chemical and biological agriculture and their respective products, you need to remember this difference and keep it prominently in mind. There needs to be effort to add content and meaning to these differences. You can't sensibly approach ecological questions until you've differentiated between technological and biological ways of thinking. If we do not carefully maintain this distinction, we are certain to end in fruitless discussions that have no hope of solving our agricultural problems and challenges.

GARDENING, FARMING, AND EATING WITHIN THE NATURAL CYCLE

Agriculture involves human intervention in and influence on the cycle of living and dead material. Thus far, our sole focus has been on manipulating the dead parts of this cycle through the expenditure of more and more energy in order to increase our food output. But if you look at human nutrition from the perspective of the cycle of living material, you will quickly realize that the currently prevailing conventional approach to food growth is effectively a form of brutal violence that we are inflicting on living systems. Doesn't the fact that the Aztecs, barefoot in their boats, were able to practice a system of agriculture that was six times as successful as our own—and more energy efficient by a factor of a thousand (see pages 139–141)—without this sort of destructive violence inspire at least a little curiosity? Is it really fair to describe this as "ignorance," "primitive culture," or "a dark age?"

It is eminently important, particularly for farmers, to know whether we are working with nonliving salt ions or with living microorganisms. The difference between these two fundamentally distinct approaches lies in properly understanding, protecting, and strengthening the syntropic, constructive force that holds the entire biosphere together and protects against destruction in the form of "everything mineralizing." The key is to recognize the amazing living power of the soil, which—even if our understanding of it is still rudimentary—keeps both dead and living material moving from organism to organism in a living cycle. I call this the power of the soil.

If we properly understand this and adjust how we think and act accordingly, then maybe we'll be getting close to something we can call truly organic!

Symbiotic Agriculture

If we are able to communicate the importance of living material and of plant, animal, and human nutrition physiology to the entire biosphere—and here to the humusphere, the narrow outermost layer of the topsoil, in particular—in a way that is believable and verifiable and to incorporate it into education for the benefit of future soil managers, then I still see hope for a nonviolent, nontoxic agricultural system, for alternative agriculture, which we could also call symbiotic agriculture.

Technologically based agriculture is unfortunately far from having anything in common with ecology. Claims to the contrary simply demonstrate a shocking level of incompetence. But what is also concerning is that so-called ecological agriculture is practically as good as technologically based itself.

Although it is unfortunately rarely noted anymore by ecologists, ecological science is fundamentally unable to challenge the chemically oriented agricultural model. It can provide practically no solid counterarguments because it operates on the basis of the same chemical NPK model of nutrition as the field it hopes to displace. This has a strong influence on ecological agriculture as a whole and is the reason why forward-looking research and practices have been so confused and ineffective. Furthermore, technological agriculture has so many economic and environmental issues it has to fight against that it can only give inadequate responses to its critics (or simply pretends they don't exist at all).

The ecological agriculture movement has always found itself on the defensive, and it still has to constantly fight for acceptance. It is still the case that barely 10 percent of European farmland is managed ecologically (with the exception of Austria, where the level is nearly 20 percent). Germany lags well behind at about 6 percent, and that is in spite of the "bio boom" of the past decade. Nobody dares to really confront the economic juggernaut that is the technological agriculture industry.

However, many years of interdisciplinary studies have convinced me that the earthworm has a real chance to defy the technologists. Fukuoka's "one straw revolution" (Fukuoka 1996, 1998,

1999, 2013) has spurred many people to think in new directions. Now someone just needs to write about the "one earthworm revolution."

Biodynamic Agriculture

There's one more theory that is worth mentioning here. It was the first explicit response to pure analytical chemistry being applied to agricultural practices. In 1924, Rudolf Steiner was prompted by concerned agronomists to put together and implement his Agriculture Course for farmers (Steiner 1979). Steiner's comprehensive philosophy and his commitment to agriculture led to the worldwide spread of biodynamic agriculture. The idea that a theory always has to be the opposite of the other shouldn't be generalized; it impedes the search for solutions. On the contrary, we need to develop methods to filter out the valuable aspects of different methods of thinking and acting and find where they intersect. This is the basis of my hope for a new, forward-thinking agricultural system.

The idea of biodynamic agriculture is as follows. Its proponents see European society as being in the midst of a five-hundred-year descent into pure, exact scientific materialism, where the only things that matter are numbers, weights, and measures (Hemleben 1981). Rudolf Steiner sought to stop this descent and ameliorate its eventual catastrophic consequences, a goal shared by modern biodynamicists as well. The nature of this descent remains the same as ever: from the collective to the individual, from systemic wholes to mere fragments. One of many examples of this:

> When it was discovered [. . .] that all organisms are made up of organs, and all organs are made up of cells—humans included—the discipline of cell pathology was born. With the help of a microscope, it was easy to see that the cells of an infected organ [. . .] took on a different configuration from those of a healthy organ. This observation prompted a change of perspective in medicine. A cell being deformed is not [. . .] the result of a disease—the disease is the result of the malformed cell.

This is a great example of how a simple change in perspective can completely change an entire area of supposedly certain scientific knowledge.

He continues:

> Unlike in ancient Greece [. . .] or Rome [. . .], doctors abandoned [. . .] a holistic perspective on the human body in favor of looking for signs of disease in individual organs [. . .]. No, now the thinking was based on individual parts, and for the most part it has remained that way. The sick man as something to be viewed "as a whole" receded into the background. Later, when bacteria and bacilli [. . .] were discovered, the perspective shifted still further. Now bacteria were seen as the causes of disease. Even today, many doctors find it unscientific and absurd to ponder the inverse perspective—could the bacteria perhaps be a side effect of sickness rather than a cause? (Hemleben 1981)

The same goes for agricultural science. I believe that the biodynamic model offers what is probably the only currently established framework for critically questioning (and ultimately solving) the problems that arise from farming and gardening being practiced exclusively in accordance with a materialistic-reductionist philosophy.

Section 3

Gardening and Farming with Humusphere

Putting it into Practice

— CHAPTER EIGHT —

Working with the Humusphere, Not Against It

From the Stone Age through Today

EXAMPLES OF SUCCESSFUL AGRICULTURAL SYSTEMS

Before I describe some encouraging projects that have already produced proof that gardening and agriculture based on the cycle of living material model can work, let's first take a little research trip.

A Paradise for Microorganisms

I have taken many journeys where I've had the good fortune of seeing some example of real nature. One such trip made an especially strong impression on my memory. It showed me that under the right conditions, it's possible to grow crops that are equal in quality to those from a primeval forest.

I was at the furthermost southern tip of Baja California, far from all road traffic and only linked to the mainland by a recently-constructed airfield. The whole thing was a broad, open bowl, scorchingly hot, a typical gigantic, now almost completely dried-out lagoon. A landscape characteristic of once water-rich Mexico. Up on the rim, the usual cactus flora proliferated. Small deposits of wind-carried sand had formed, and the soil was composed of fully eroded clay-like sand, sparkling with millions of silicate granules. It was only a few steps from the cactus formation to the semi-desert and desert.

Directly under the bowl's rim were rows of corn, rather meager alfalfa, humble and very dry wheat fields. Toward the middle, the soil became noticeably better, and the cactus shrubland stopped at the uncultivated rows. In the deeply sunken middle, however, dark greenery rose up like a primeval forest. There was no doubt that this spot represented the remains of the erstwhile lagoon.

It was a paradise. The paradise was made up of old mango and apricot trees and sugarcane. Tiny green islands were connected by narrow, clear streams with emerald algae pads floating on them and banks brimming with fresh pondweed.

It was explained to me that the sugarcane was cut two to three times per year, but other than that nothing was done. Nobody tended this black, superbly loose soil, which lay there as soft and airy as a sponge. The sugarcane stalks grew four meters tall and almost as thick as your arm, with entire sheaves of young, natural tillers. Ripe fruits rained down from the gigantic, darkly-shadowed mango trees, not picked by anyone at the time. The soil was coated with a layer, half a meter high in some places, of plant remains, rotting fruits, twigs, dry leaves, and stalks. I was struck by this "global cellulose capital," as I like to call it, which we senselessly squander in our gardens and fields.

After a plant's death, its cellulose is a gold mine, one that was not valued when it was present all over in nature. But today it is no longer present in such quantities. [. . .] Biological tests on the incomparable natural humus found in Baja California revealed a correspondingly great abundance of organic material, swarming with countless organisms. Every part of the edaphon was represented. This soil was alive down to the smallest crumb. It displayed a harmoniously ordered system of life, in which breakdown was immediately followed by building up again. It was a single huge oligosaprobic biocenosis, with everything in the pH range of 7–7.5. [. . .] Embedded in this miniature paradise were individual silicate granules (whereas the soil above the bowl's rim consisted solely of eroded silicates), but even they were fully encrusted by colonies of soil algae and lithiobiontic decomposers.

Tests revealed that the ratios of the various groups of organisms to each other were close to ideal levels, which are as follows:

Bacteria	27%
Soil algae	20%
Diatoms	9%
Soil fungi	9%
Rhizopods and amoeba	11%
Infusoria (flagellates and ciliates)	6%
Rotifers	3%
Nematodes	2%
Indigenous microflora	8%
Cysts and spores	5%
Soil life altogether	= 100%

All humus production should aim for these proportions!

Human settlement and agriculture have changed the life in the soil. Fertilizing with stable manure and slurry, as well as with ammonia salts, always increases the number of bacteria. This explains why you can test a gram of good soil many times and come up with one to five billion bacteria, but only 50,000 to 100,000 soil algae and only 10,000 protozoa. The polysaprobic and putrefactive bacteria that enter the soil every time that stable manure is applied to it survive for a long time due to their resistances. Nowadays they are predominant in all cultivated soil and integrate only with difficulty into other biocenosises once theirs breaks down. They are always accompanied by large numbers of mostly facultative anaerobic bacteria, whose function is not yet fully known [. . .]" (Francé-Harrar 1957, 68–70)

So reports Annie Francé-Harrar. If this biological potential is fully realized, then, in contrast to our modern treatment of the soil with its destructive side effects, it's conceivable that we may once again be able to achieve harvests five times as large as our current ones!

Successful Agricultural Systems From Then through Now

Essentially every advanced culture achieved agricultural harvest results that we are unable to replicate with any of our modern methods. I believe we should focus our attention on what role microorganism ecology and related research is going to play in the future of agriculture. Back in 1995, over 100,000 metric tons of microorganisms were already successfully being used as "living fertilizer." And we shouldn't ignore the practical agricultural and gardening experiments presented in the last chapter (starting on page 119), which are both useful to us in and of themselves and also prove that they really do bring higher yields and, above all, better toxin-free harvests.

But this book also addresses the larger question of whether we are going to be able to feed a global population of seven (or one day, perhaps even ten billion) people. I hope that the following examples will provide some hope that a solution is indeed possible.

The Nile Civilizations

The chemical composition of the material suspended in the Nile silt (see page 137 of the color images) has been analyzed many times. But you rarely or never read about attempts to investigate the yet unknown potential benefits that this silty, oxygen-rich, sun-warmed water may offer to plankton and edaphon growth. Annie Francé-Harrar (1957, 128), however, reports the following biological analysis:

The floating gardens must have looked like this, or at least similar.

Our ancestors already recognized the value of the nutrient-rich Nile silt a thousand years ago.

Rice yields are improved considerably by the presence of rice crabs.

Excerpt from a Research Paper

(Nile silt from flooded fields in Badrashin near Saqqara, 10/21/1936)

Structure: loose, almost completely composed of organic-inorganic zoogloea Mineral granules: almost exclusively silicates, many feldspars (plagioclases), pH value of 7

Organisms contained in 1 cubic centimeter of silt sediment:

- 92,000 protozoa
- 75,000,000 indigenous microflora
- 2,000,000 soil fungi, spores, and cysts
- 150,000,000 bacteria
- 227,092,000

Preponderance of lithobionts and bacteria (oligosaprobes dominant).

The favorable growing conditions caused by the annual flooding are also accompanied by a massive increase in the quantity of microbial organisms. Even in ancient times, these organisms were immediately and freshly available to crops. The microorganisms themselves had ample access to true mineral rock dust particles, the inorganic material component of the Nile silt. If you imagine a densely woven symbiosis between rock dust, oxygen-rich water, sun, warmth, plankton, edaphon, and plant root endocytosis, the "now unattainable" harvest yields of the time seem realistic.

Historians have discovered accounts of earthworms that show that the Egyptians had some awareness of the function of the edaphon: "Cleopatra [. . .] recognized the earthworm's contribution to the Egyptian agriculture by declaring this animal sacred. Egyptians were not allowed to remove so much as a single worm from the land of Egypt, and even farmers were not allowed to touch an earthworm for fear of offending the god of fertility" (Minnich 1977, 66).

I am convinced that all of the Urstrom cultures, as well as the terrace cultures, had a close connection to the water plankton and to plant-protein nutrition.

The Aztecs

Around AD 1300, a group of Indians arrived in modern Mexico and founded a city on a marshy island in the middle of Lake Texcoco. Over the course of the next two hundred years, this city became the center of the Aztec Empire, which we continue to investigate today with ever-increasing admiration as the "Venice of the Americas." This center, Tenochtitlan, grew into a mighty capital with three hundred thousand inhabitants, "four times as many as lived in London under Henry VIII at the same time." (Author's note: the city's ruins have been almost completely overbuilt by the modern-day Mexican capital, Mexico City.)

This population was primarily fed through a system of agriculture that, according to current research, began with floating woven baskets. The baskets were filled with plant residue and silt scooped into the baskets from the shallow lake, and beans and corn were grown in these floating plant beds. Over time, these "floating gardens" (Francé-Harrar 1957) became islands, which were anchored to the lake floor with willow trees planted around the edges. These newly created islands ultimately formed a network of channels and were worked with boats and human power, leading to the most successful agricultural system ever known (compare pages 135 and 140 of the color images). "The Aztecs regarded lakes and rivers as gods, and were duly blessed with fertile soil of a quality that we have never equaled since, not even with the most modern agricultural methods" (NRK video 1997).

Lake Texcoco, which was very large at the time, served as a food source for up to fifty different species of migratory birds. The lake must have been very rich in nutrients to have been able to supply food—plankton—for so many birds. Some archaeologists believe that the Aztecs developed cannibalistic practices due to a lack of protein, but others reject the idea. What is clear, however, is that they used finely woven nets to fish for large quantities of very small aquatic insects and plankton in the lake after the birds had migrated onward. They used them to make a very nutritious protein paste that they considered a delicacy.

Report on Aztec agriculture in Lake Texcoco.

Nowadays, the spirulina alga has become famous, is a common subject of conversation, and can be found in many stores. "Spirulina, a nutrient-rich alga, once an important natural resource in the diet of the Aztecs, could feed many of the world's starving people" (Furst 1978). "The Aztecs in Mexico and the Africans living around Lake Chad both independently discovered the nutritional potential of the blue-green spirulina algae. [. . .] Spirulina produces harvests with dry weights around 12.5 times higher than wheat per hectare per year, and 8.3 times higher than soybeans. The corresponding protein yield in spirulina is 70 times greater than in wheat and 14.6 times greater than in soybeans. Spirulina's protein yield per hectare per year is 78.6 times greater than that of meat fed with wheat-based feed, and 27.8 times greater than that of meat fed with green fodder" (Hubendick 1988, 168).

Given the knowledge that plants directly absorb cytoplasm as a nutrient source, it is logical to investigate whether cytoplasm can be found in areas where many plants flourish. It was indeed found in Aztec regions. And I'd like to point out that since the Aztecs used silt from Lake Texcoco as the basis of their successful agricultural methods, the unusually rich and diverse plankton naturally also functioned as a direct nutrient source for the plants!

Rice Crab Agriculture

A video from my archive (Sakura 1984) shows images of rice cultivation, grown hydroculturally over endless arduously constructed, unbroken terraces (compare page 137 of the color images). This shallow, almost motionless, sunlit water is home to rice crabs. Their presence guarantees a successful harvest. (Similar information can be found in King 1911, with new editions published 1984 and 2004.) The crabs swim and scuttle over the loamy ground in the flooded rice paddies, continuously stirring up soil particles. This means that they are constantly mixing parts of the silty soil with the water, which prevents weed seeds from settling and germinating. The rice plants themselves are planted firmly enough to be unaffected. This movement of water and silt continuously taking place around them prepares a biotope that is favorable for plankton growth in the lightly sunlit, lukewarm water, which is filled with true mineral particles and floating organic material. This plankton provides sustenance for the rice crabs, the rice plants, and the Chinese. At the same time, the kilometer-long planted border areas provide the necessary reinforcement for the entire terrace construction with their roots, with the aboveground parts of the plants constantly collecting solar energy and their roots in turn feeding off the edaphon in the soil. Artificial fertilizers would destroy all that.

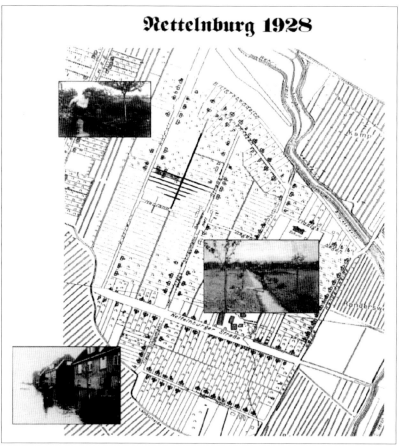

Nettelnburg, located in the western marshes of the Elbe River, my original hometown. My grandparents carried out work on a pioneering settlement plan here. Food was largely provided through self-sufficient means in the form of vegetables, potatoes, fruit, and small livestock. Fostering and maintaining the edaphon as the basis for this system was indispensable.

The Elbe Valley around Hamburg

The Elbe Valley around Hamburg is a 10-kilometer-wide remnant of the last ice age and is filled with solid loam. The unusually high fertility of this area is attributed to this loam. The entire region is densely permeated by ditches and canals, and it is strictly controlled by locks that are closed when the Elbe floods and opened to allow water to drain.

In one of these ditches at my grandparents' place in the Nettelnburg section of Bergedorf, I caught my tadpoles, water fleas, and amoeba. My grandparents had resettled from the city of Hamburg to 1,000 square meters of Elbe Valley loam, in an area with the descriptive name of Nettelnburg (derived from *Nessel*, German for "nettle," a plant that flourishes especially well in the region's soil), in 1928. It wasn't until fifty years later that I understood that this area's fertility was in all probability based more on the large quantities of plankton, green plant matter, and root material that was produced each year in the kilometer-long ditches than on the loam alone. I also remember that the spillover from our cesspool emptied into one of these ditches. I always found the best rabbit food there, but the ditch always smelled like sewage within a few meters of the cesspool until the smell transitioned into a regular earthen smell and the smell of grass and ground ivy (*Glechoma hederacea*).

The cleaning and care of the whole network of ditches was strictly regulated, and it regularly provided us with a valuable supply of silt, aquatic plants, grass, and wild herbs as well as roots, plankton, and edaphon that we could directly add to our vegetable and fruit garden. Today, the ditches have been converted into pipes.

The Hunza

The Hunza people of the Himalayan Mountains continue to practice a form of agriculture that should command great admiration and respect, even though from a purely technological point of view it is it the simplest system possible (see next page).

According to Hunza agricultural tradition, the gray-black piles of rock mate with the snow-white glaciers high up in the Himalayas, producing new glaciers and new rock. The Hunza are abso-

The residents of the Hunza Valley make use of a nutrient-rich, fertile hydrocultural system.

The gardens of the Inca were another form of hydroculture.

lutely dependent on these glaciers. They are the source of glacial milk, melted water with a milk-like murkiness, which is made up of countless microscopic rock particles that are rubbed off the rock by the glaciers, which weigh many metric tons. These are true minerals (i.e., rock dust in its very finest form, which contains essentially every basic element that exists on the planet). In terms of endocytosis, this means that plants can directly absorb these rock particles into their cytoplasm through the cell walls of their root hairs and process them there as needed.

But beneath the mighty Himalayan glaciers is nothing but rocky desert—there are no plants. To deal with this, the Hunza built "gravity canals," some of which are kilometers in length, into the overhanging rock face, through which the glacial milk is channeled 1,000 meters onto an originally barren rock plateau. These narrow canals, powered only by gravity, guide the rock dust–containing water. The slope needs to be precisely calibrated: if too much water flows down too fast, it will break apart the canal walls, which are only sealed with loam; and if too little water flows down too slowly, the particles suspended in it will settle as silt too early, clogging up the whole system. The fact that it works—and with a length of many kilometers over an elevation difference of 1,000 meters—is doubtless a sign of skillful engineering. On the barren rock plateau, the Hunza have set up an entirely artificial farming area and created a "Garden of Eden" that sustains the entire population. This form of agriculture also employs the principles described in this book, as the plankton and edaphon in the oxygen-rich, gently sunlit, mineral-containing glacial water forms the living foundation for the plants and was used by people in previous eras in a way that has become utterly foreign to our modern technological conception of the world.

In all of the forms of hydroculture mentioned so far, we find huge amounts of living cytoplasm as well as proteins and plankton, and we always find especially favorable conditions for them to multiply. Mountain streams carry glacial milk or volcanic ash with real mineral particles in the form of oxygen-rich water through long, gently flowing channels to slowly flowing, shallow, warm lagoons. They contain ideal, abundantly multiplying populations of so-called infusoria, whose living cytoplasm is absorbed directly as food by plants via endocytosis. The plants draw the cytoplasm directly out of the tiny creatures, as described by Hamaker (Tomkins and Bird 1998, 188). Just as saltwater plankton is known to form the basis of all further life in the oceans, I believe that freshwater plankton formed the basis for all hydrocultures. This substantially high amount of edaphic organisms and the equally high amount of organic material in the original soil species available as food for this edaphon (Francé-Harrar 1957) suggests that we should investigate the capacity and potential of these "old" hydrocultural systems for the growth of water plankton as direct, living food for plants.

There are countless other examples where true mineral material in the form of glacial milk or volcanic or eroded dust, organic nutrients, microorganisms, water, oxygen, light, and warmth have led to excellent harvests. Kilometers of drainage canals or irrigation systems, annual flooding, and silt supply the land and the plants growing on it with greater quantities of plankton than has yet been realized, including all of its constituent protein. Rice crabs in China are a sure sign of a good rice harvest, and the Aztec floating gardens were teeming with spirulina algae. Stinging nettle liquid is the purest infusoria culture, and Francé-Harrar's edaphon seeding, Rusch's Symbioflor, and the effective microorganisms (or preparations of them) made famous by Higa, along with all other microbial preparations, build on the same basic foundation. *But no researcher or institution has ever researched the process of plant endocytosis in connection with this.* Each of these methods has only been analyzed chemically, in terms of their levels of the basic elements nitrogen, phosphorus, and potassium.

The Rye Finns

"A people emigrates, with seed grains in a glove." It took me three years to discover what the basis of this saying was. The glove was a bird skin, probably from an auk (*Alcidae*), that contained about half a liter of seed grains. To emigrate with one of these was certainly not beyond an entire group of people. This was 350 years ago, and it was the Rye Finns who were leaving their homeland. They wandered around the entire Baltic Sea before finally settling near Oslo, where their descendants still live in Finnskogen, "the forest of the Finns." These Finns employed a system of agriculture that they called *huuhta*, or "slash-and-burn." On a relatively small scale during the course of constant migration through long stretches of what was then almost endless forest, the damage this caused was probably limited, and the forest probably returned to its previous state a short time after the clearing and a brief period of growing.

A Norwegian anthropologist, Per Martin Tvengsberg, reconstructed a model of huuhta on a small scale. He wrote the following to me: "The forest rye produced up to 12,000 grains from a single seed grain. My cultivation in an artifically-set-up [*sic*] huuhta system, which I've been studying since 1986, showed that one seed grain produces a yield of 12,000 grains. Each seed grain turned into a large plant with 150 to 180 stalks 2.3 meters in height, with 60 to 80 grains on each stalk." Siegfried Lange (Lau 1990 and 2017) was able to reach up to sixty-thousand grains per seed grain using this method (see the top of page 64 of the color images).

The Rye Finns' Method Today

Here is a report from someone who experimented with the Rye Finns' method:

How to harvest at least 10,000 grains from a single grain of rye:

Take good, germinable rye grains and place them individually in the soil or in pots at 50-centimeter intervals two to three weeks after the summer solstice. If using pots, however, it is essential to remove the plants after about three weeks or else they will usually grow too large for them.

The grain will tiller unusually strongly from the heat of the summer until well into the winter, and it is quite capable of surviving the winter months. You can count the first stalks by early May. There should be at least a hundred stalks per grain. (My best plant grew precisely 196 ears.)

Interval: 50 cm

Sowing period: Late June/early July (or several weeks earlier in Scandinavia)

Harvest: August of the following year

Potential yield per hectare:

- 1 grain provides 100 ears (or even up to 200)
- 1 ear contains about 100 grains
- 1 plant thus has 100 x 100 = 10,000 grains
- 4 plants per square meter produce 40,000 grains (with a 1,000-kernel weight, or TKW, of approximately 50 grams)
- 40,000 grains thus yield 2 kg per square meter
- Per hectare, that comes out to 10,000 m2 x 2 kg = 20,000 kg or 20 metric tons

But since the grain yield per plant can be even higher than the value given here, this in fact represents a poor harvest!"

(Lange 1996; see also the top of page 64 of the color images)

This puts us on the verge of the "impossible"! What could account for these results? The Finns' auk gloves? Or the sowing period? What are we today to make of what was happening in the Scandinavian forests two hundred years before Justus von Liebig applied "chemistry" to agriculture? Of course, what you always hear is that the grains can grow so well because the ashes of burned forests contain so many water-soluble "nutrient salts." If this is the explanation, then the question is: Why today, 170 years after Liebig, is nobody able to achieve this sort of yield while we're downright swimming in NPK fertilizers?

Siegfried Lange's "secret" lies in his use of wood shavings. His garden had an area of 600 square meters. In this much area, he used 25 cubic meters of shavings over the course of three years: wood shavings and sawdust, but no particle board, and nothing with waterproofing, coloring, glue, or formalin. "In my experience, bark of any variety is even more useful than shavings," he wrote to me. Essentially, both are the forest in finely shredded form.

He collected his shavings between winter and spring, then stored them in piles of about 1 meter in height under the open air. Over the course of the summer, he kept the piles lightly aerated, which isn't the easiest thing to do. Better, as he discovered, is to mix one part shavings with one part slurry liquid. Soil can also be mixed in to provide up to 10 percent of the volume. Slurry and soil are full of microorganisms. Bark should be handled similarly, but it needs to be stored for two full years.

In a certain sense, what is happening here is pre-composting, microbially set in motion with the help of the soil and the slurry. The process is slower in wood than in regular compost, meaning that less energy and fewer nutrients are lost, and can presumably instead be utilized as living cytoplasm. After one or two summers, the mixture is then applied as mulch in a 10-centimeter layer.

My skepticism when it comes to the ash theory and the NPK theory in general has probably helped me find a possible alternative explanation: plankton, edaphon, microorganisms. The forest has spent centuries building up raw humus: acidic, almost fermented foliage, needles, and wood. Could it be that with the alkaline ashes, microbial growth was set in motion that helped the Rye Finns, at least for one growing period, to attain such unbelievable grain yields? "In the following year, they would plant only turnips, and in the third year only grass would grow, so they would keep moving in search of a new suitable patch of forest" (Tvengsberg 1992). When I suggested the possibility of invigorating the soil to the anthropologist Tvengsberg, he was reminded of unusual microbial activity in "old" forest litter.

It is natural to assume here that it is once again the edaphon or some other force that is creating fertilizer from the wood-shaving piles. After all, you cannot achieve the results described with either NPK fertilizers or with "finished" compost. It can only be living edaphon in full vitality.

The Garden Estate

The basic concept behind the modern phenomenon of urban gardening had reached us in Norway by the 1980s in the form of the first permaculture seminars and the garden-estate model. Allow me to briefly address the relevant history. After World War II and the subsequent forced expulsions of ethnic Germans from the Baltic countries, the farmer and former major landowner Wolfgang von Haller began to carve out a new, ecologically friendly existence in an unkempt garden. Given the garden and kitchen work that was necessary at the time, and armed with a clear understanding of the horrific destruction of the world (not only due to the war), he set a goal of "growing things naturally and communicating ecological discoveries." In 1949, he founded the Gesellschaft Boden und Gesundheit (Society for Soil and Health) with his brother Albert von Haller and other kindred spirits, which was later based in Langenburg, Baden-Württemberg up until its dissolution. Among many other ideas and practical enterprises, the organization absorbed the Gärtnerhof-Gesellschaft (Estate Garden Society), which had been founded in 1946.

Over the following thirty years, Gartenstraße (Garden Street) in the town of Langenburg became an ecological center. Starting in 1962 at the latest, seminars were held on practical ecology, gardening, and agriculture, often in conjunction with Bauernschule Hohenlohe (Hohenlohe Agricultural School) in Weckelweiler, which had also become famous for its strong commitment to organic agriculture. (Author's note: another pioneer of organic agriculture and friend of the garden-estate concept, principal Fritz Strempfer, was active here for decades.) From 1954, Wolfgang von Haller and the Gesellschaft Boden und Gesundheit regularly issued a magazine of the same name, with the subtitle *Journal of Applied Ecology*. (Author's note: it ultimately merged with the magazine *Natürlich Gärtnern & Anders Leben* [*Gardening Naturally and Living Differently*].) They also published over seventy special editions (author's note: some of these are available through the OLV Organischer Landbau Verlag, Kevelaer), which mostly drew from the magazines, along with a wide variety of other articles and books. (Author's note: these publications reached a large audience for the time, even beyond the borders of what was then West Germany.)

The historic garden-estate concept of Leberecht Migge (1918) and Max Karl Schwarz (1974) was later further developed by Per Krusche, Dirk Althaus, and Ingo Gabriele (1982) and applied to permaculture by Margit and Declan Kennedy and by Oswald Hitschfeld (2003) in *Der Kleinsthof und andere gärtnerisch-landwirtschaftliche Nebenerwerbsstellen: Ein sicherer Weg aus der Krise* (*The Small Estate and Other Gardening and Agricultural Smallholdings: A Safe Way out of the Crisis*). In the foreword to this work, Kurt Walter Lau writes: "It's possible that, someday not too terribly far in the future, smallholdings will represent a sort of rescue station for some of our fellow men. Perhaps a sort of ark in the chaotic storm of our times from which a new way of thinking and acting will spread forth [. . .]."

The best summary of all of these ideas and undertakings I have found was written in 1978 by Maria and Per Krusche. In Per Krusche's garden estate, a wide variety of the different functions and daily tasks associated with technological methods are broken down into smaller units and grouped together in symbiotic communities. He labels a drawing signed "Maria/Per-78" with the following:

> Scaled comparison of cultivation and growing area: 120 residential units with about 600 inhabitants that are only set up for self-sufficiency to a limited degree can be supplied by about six hectares (250 x 250 meters) of garden estates (without grains and milk). If the land is used entirely for grains, 40 residential units with 200 inhabitants can be supplied. For a milk supply, each cow requires about one hectare to produce an average of 10

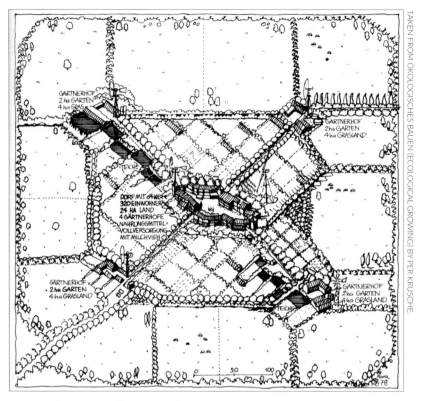

Four garden estates (one in each corner, each with two hectares of garden land and four hectares of grassland) flank an eco-village with 64 residential units and 320 inhabitants on 24 hectares of land in order to supply it with food, including milk.

liters of milk per day over the course of a year. This means that about 16 households with 80 people could be fully provisioned, including milk (four cows), by six hectares. With effective storage, this direct allocation removes the need for a significant proportion of the effort and land area devoted to transportation, processing, packaging, distribution, and waste disposal.

The system also provides a significant quality of life increase without having to sacrifice the advantages presented by the density of cities. If additional intensive food production methods also integrated into the system, such as fish farming, greenhouse cultivation, mushrooms, small animals, or vines and trees that provide fruit, the amount of area needed is reduced significantly further still. The presence of various different habitats connected by passageways also increases and maintains the level of biological diversity.

Not Just Talking, But Doing: The Erga Gard Partial Garden Estate in Norway

Ingvald Erga scaled Per Krusche's aforementioned 24-hectare model to fit a 7-hectare piece of land and is slowly but surely continuing to develop it. Years of experience have been and continue to be put into the establishment of this eco-estate:

Protection against Inclement Weather

For centuries, the inhabitants of Jæren (a coastal region of southern Norway) have known how to protect themselves against wind and weather. In the past, partially buried houses were common. The house in the Erga Gard is partially buried, and as a result it is protected from cold winds coming from the northwest and the east. The kilometers of windbreak plants around the 7-hectare estate consist of over thirty different flowering and fruit-bearing trees and shrubs, which serve as food sources for insects and birds.

Passive Solar Heating

The sun's heat is absorbed into the heavy building materials via thermal energy storage—a millennia-old method. This is effectively taken advantage of in the house at the Erga Gard, which has as many windows as possible facing the south and a heavy concrete wall serving as the rear wall of the living room. The same approach was employed in the greenhouse, where heat is absorbed by the rugged, 2-meter-thick wall of boulders and can be released again at night.

Ecologically Friendly Food Production

The diverse and labor-intensive system of agriculture follows the principle of growing various things in small plots.

Reduced Land Use

According to the Club of Rome, 2,000–4,000 square meters of land is necessary to feed one person under conventional industrial agriculture. At the Erga Gard, 325 square meters of professionally managed ecological production is enough to do the same (as has been comprehensively proven and documented).

Erga Gard also offers courses and seminars on subjects like ecological gardening and tea cultivation, healthy living, alternative energy, and permaculture. As of 2012, there were also plans for a rest area for cross-country bicyclists traveling the North Sea Cycle Route, for a migrant bird observatory in a pavilion among reeds, and for the establishment of a residential community.

Models like this one can also incidentally offer new ways of living and interacting that foster social cohesion and can create new jobs due to the labor-intensive growing methods. The latter advantage might even be applicable to a certain extent to people who are lacking qualifications, a group that is being pushed more and more out of the job market and into the fringes of society. The system's direct coupling of various different aspects of life—such as living spaces, free time, ecological waste utilization, work, and education—with tangible output in the form of food and social support could be the starting point for a new social consciousness among tomorrow's generation.

But ideas like the Erga Gard won't happen until many more people change their way of thinking.

Urban Gardening and Community Gardens

But we don't have to go so far as to build completely new settlements to open people up to new, healthy forms of gardening. On a small scale, it's even happening in cities. Growing cities, growing food: urban agriculture, well known from the war years to those of us of a certain age and a curiosity or challenge to the younger generations. *Die Wiederkehr der Gärten. Kleinlandwirtschaft im Zeitalter der Globalisierung (The Return of the Garden: Small-Scale Agriculture in the Age of Globalization)*, by Elisabeth Meyer-Renschhausen and Anne Holl (2000), which was presented simply as a vision, has happily been overtaken by reality in the form of the numerous urban gardening projects that have sprung up in both big cities and smaller towns like mushrooms out of the ground.

Large-Scale Compost Plots

And now a completely different sort of example: *organopóniocs*, large-scale compost plots in Havana, Cuba (Hoffmann 1999). These are intensive gardens that line Havana's main arterial roads. They are planted on land that was previously unusable for agricultural purposes. Concrete or prefabricated tubs 1 meter in width, 30 centimeters in depth, and 100 meters in length are filled with a 1:1 mixture of soil and organic material, with no further fertilization.

The following is quoted from a report by Elisa Claro from 1999: "Four years ago, the Valero family started with a small parcel, and today they run one of the largest so-called *organopónicos* (vegetable gardens) in the community. 'Anything that is planted from the heart yields a good harvest' is their motto, and indeed, the Valeros far exceed the national average of 20 kilograms per square meter under cultivation at 25.79 [. . .]."

Fish Remains as Fertilizer

In the tiny South Atlantic island of Tristan da Cunha between Africa and South America, fish and crab remains are still laid out alternatingly with potato plants, much like how Norwegians used to *la eit sildahove attmed kvart froeple* (place a herring at the side of every potato plant; see next page). There are also similar reports of traditional methods employing fish remains as plant fertilizers among the Indian tribes of North America.

CREATING THE BEST POSSIBLE LIVING CONDITIONS FOR THE EDAPHON

So how do we create the best conditions for the edaphon (and thus the most edaphon), the creators of our field and garden soil's structure and a living food source for our crops? The edaphon needs air, food, moisture, warmth, room to develop, and protec-

A traditional Norwegian fertilization method using fish remains. Bottom: Potato harvest from the plot. This method is not suitable for urban or suburban gardens.

tion against wind and weather. Since all of this takes place in a constant, continuously repeating cycle, the process can be started at any arbitrary point. Sooner or later you always end up back where you started, as long as you stick with it.

There are countless examples of how you can and should care for your garden. I have tried to compile the places where a gardener can provide the edaphon with food. You will be surprised how many possibilities there are. Rainwater barrels colonized by algae, liquid fertilizer made from stinging nettle filled with plankton, and a nutrient-rich cover over the soil are already a good start.

Our garden soil is coated with a thin, almost invisible lawn made up of green algae. Earthworms (among others) "graze" on this field of algae at night. They also collect loose plant parts and consolidate them into small "compost piles" above their burrows. Anyone who has recognized these interrelationships cannot help but recognize that this diversity of life, this perfect interplay between the forces of nature, must not simply be thoughtlessly annihilated with artificial mineral fertilizers and pesticides!

My Appeal to Educators and Scientists

Please devote yourself to this subject. I call on agricultural researchers, teachers at technical and agricultural schools, all organizations and representatives of farmers: take guidance from the many "old" models presented in the previous section, as well as from the ever-increasing number of practical examples of proven, functional undertakings that have already carried these methods over into modern agriculture. Research intensively in this area so that our soil, our plants, our healthy foodstuffs, and not least our farmers and consumers—so that all of us—can have a future.

— CHAPTER NINE —

Practical Examples and 'Recipes'

One of my major objectives is to establish this subject in schools and kindergartens, but also in extracurricular adult education. If you, dear reader, have become curious after reading the last chapter and want to try out agricultural methods that follow the principle of the cycle of living material yourself, then you can find accounts of my own experiences, practical tips, and "recipes" here. I wish you all the best in your gardening endeavors and rich harvests!

The examples described here are old, but at the same time very new gardening methods, and are equally applicable to farming. All of these examples are documented in detail in further writings and articles as well as in my Little Humusphere Museum in Norway, Jochen Koller's Humuseum in the Allgäu, and the Gut-Neuenhof-Stiftung along the lower Rhine (see appendix for addresses). They need to be developed further to the point where they are also transferrable to and manageable over larger areas. Then they'll be able to change the entire structure of farming and the way we grow our food. This represents a massive research and development opportunity!

GOOD OLD COMPOST

We need to completely change our view of compost. Traditionally, self-sufficient gardeners have collected all the food for the life in the soil in the form of a compost pile. And at first, so much is going on within it, so turbulently, that nothing at all can sprout or grow there. Later, once the compost pile has ripened and calmed down, it becomes mild and is mixed with sand and used for seeding and—in a more highly concentrated form—during the first propagation phase. Diligently and persistently applied, ripe compost can support nearly disease-free growth (Chaboussou 1985; Howard 1943, 1979, 2005; King 1911, 1984, 2005; Seifert 1948, 1991; Sheffield 1949, 1970, 2016; von Haller and von Haller 1978). But when growing in pots, we sometimes find that ripe compost can reach the limit of its energy reserves.

In fact, compost does not carry enough reserves to provide the energy and nutrients that small plants need to grow within the limited soil volume in plant containers. Indeed, it has already squandered this energy in its shady spot at the back of the garden during the initial hectic phase (see the upper left of page 71 of the color images). Thus, composting all the way until ripe compost is produced does indeed produce mild, healthy soil, but it is missing important organic recomposition processes and energy that have already taken place or been expended during the ripening process.

That's why a basic tenet of organic agriculture is: provide the freshest, most nutrient-rich organic material as food for the edaphon at the location where the soil life is closest to the plant roots (i.e., as a soil cover around the plants). The function of ripe compost can then, at least in part, be taken over by well-nourished garden soil. This is, after the idea of the cycle of living material, the most important aspect of the organic-biological movement that Müller-Rusch recommends for any agricultural method.

Now we go from composting to the next method: mulching.

GREEN MANURE? MULCH!

The function of green manure spread over a field has long been known and acknowledged: the accrued plants are left on the field and are—while fresh—plowed in or mulched for the purpose of "soil improvement." Really, we've always known what's good for the soil. Cover the soil loosely and airily with all of the fresh plant material that accrues in the garden, and make a feed-mulch cover out of it for your edaphon. It should be immediately clear that the key here is not the NPK content of the material but the quality and freshness of the plant remains used. Every green plant is a collector of sunlight and suitable for use as green manure—including weeds!

A few more thoughts on the conventional methods of mulching: It's clear again here how heavily the theory one ascribes to determines the further procedure. If you're guided by the salt ion theory, you'll be mulching with black plastic. But if you're guided by the life in the soil, the edaphon, then you'll quickly recognize that without holes in the plastic, the plant roots won't receive enough water or oxygen. The stale, carbon dioxide–containing air breathed by the edaphon cannot make it out of the humus layer and instead is kept there, poisoning the edaphon and the plant roots. It also gets very hot directly under the plastic, reducing the amount of edaphon.

If you mulch with waste material (e.g., with nongreen leaves or straw), it will simply remain as waste, with limited nutritional value, as the leaves have already given most of their energy back to the tree, the straw back to the grain, and excrement is nothing more than the leftovers that animals and people were unable to make use of in their bodies. But it is nonetheless better than plastic, and it does encourage edaphon growth, just not as well as it could. Think about life and how it is maintained. If you have a cow, a sheep, or a guinea pig, you don't just toss it a single bale of straw or handful of woodchips each year and expect a sprightly guinea pig, twin lambs, or 6,000 liters of milk. You have to supply your animals with the foodstuffs they need on a very regular basis. The same thing goes for the edaphon!

The edaphon can also be fed with hay, twigs, and leaves as well as limited amounts of wood and straw. The fresh green parts are rich in chlorophyll and living cytoplasm, but the supposedly "dead" parts (like wood) are also "eaten" by microorganisms—usually more slowly—and are fully and completely incorporated into the living organism. However, you must take care to mix in the proper ratio of fresh material and to store it for the right amount of time.

Mulch Is a Complete Habitat for the Edaphon

Or to put it another way: a climate-regulated nutrient layer for just about all inhabitants of the soil. The fresher and richer in nutrients it is, the more vital and lower in contaminants, the better.

Åkerstedt's Grass Mulch Method

The Swede Nils Åkerstedt has come up with a systematic method of mulching. He plants everything, from flowers to vegetables, in pure sand with a cover of grass and herbs applied twice per growing period—with success. Since this method works in sterile sand, it can be used even if you have only sandy soil available or if you need an emergency solution due to soil sickness. It's true that Åkerstedt believes that the effect lies in the "mineralization" of the grass. But I think that this layer of fresh grass, which Åkerstedt's instructions say to keep airy and always slightly moist, is an excellent microorganism culture that finds its way to the plant roots in the sand on its own. The airy, moist grass and the countless microorganisms multiplying within it are direct relatives of the soil flora (i.e., the edaphon).

If this grass mulch or fresh-grass method is responsive in pure sand—that is, even without soil around the roots of the planted seedling—then this can be taken as an indication confirming Schanderl's remutation model, according to which roots, dug-in plant remains, and all seeds (once they begin to germinate) "send out," in a sense, the things living inside them, the endophytes, to find nutrients and to produce them by way of their own reproduction.

Nils Åkerstedt's *The Book of Mulch—and on Gardening in the Sand* (1993) is available only in Swedish, sadly. It is the first and only gardening book that comprehensibly details how to cause things to grow with only green grass and sand. On page 8, Åkerstedt writes the following under the heading "simplified growing method":

> It may sound unbelievable that you can replace any of the types of fertilizer sold in stores with simple grass or finely cut green plants. But to all those who doubt whether they should switch over to this method of fertilizing and growing, I propose a small experiment—and why shouldn't we call this "research in simplified growing methods?" Take two small plants, e.g. geraniums, petunias, or marigolds, and place one of them in a flower pot with regularly commercially-available soil and the other one in a pot with pure sand. The plants should have a small clump of dirt around their roots when you place them in the pots. Water the plant in the potting soil with conventional fertilizer, but only provide the other plant, the one in the pure sand, with a three to four centimeter layer of fresh, cut-up grass. Place a new layer of grass on top of the previous one every month.

However, you must make sure that the grass layer never completely dries out. The sand should be coarse, like the type used in bricklaying.

This simple description inspired me to try it out myself. For my experiment, I used half-size lettuce from "Grüner Stern" asparagus salad. After one week, the plants already exhibited healthy, vigorous growth. Its superiority to the control plant in the conventional soil was shocking! Nils Åkerstedt's presumption is that the plants absorb the elements nitrogen, phosphorus, and potassium from the grass mulch in forms that they are able to make use of.

The small clumps of dirt that Åkerstedt expressly recommends the plants' roots have around them offers a very gentle way of

Top: Grass mulch plot with cauliflower saplings. Left and right: Harvest from the grass mulch plot.

Åkerstedt's grass mulch method has already been tested on larger land areas.

bringing over roots, root mucosa, the microorganisms living within it, and a portion of the edaphon into the otherwise "pure" sand. Further food comes in the form of infusoria from above, like in a hay infusion, which continuously multiply in the lightly moist, airy grass.

But that isn't quite everything. Åkerstedt also stores his green-plant mass for a time: he mixes together 40 percent finely cut grass and weeds, 40 percent sand, and 20 percent peat-free potting substrate by hand or with a machine, then seals it in an airtight black plastic sack or a bin, or piles the mixture in large quantities like a compost pile. Then the whole thing, moderately moist and packed together as tightly as possible, is stored out in the open and covered with a fabric so that flies can't lay their eggs in it. After two months at the earliest, but also after being kept frost-free over the winter, the containers or sacks are opened and aired out for at least a week. The mixed-in sand keeps the mass from becoming matted together. Åkerstedt uses this mixture "like a natural fertilizer." To make potting soil from it, he mixes together 30–40 percent grass-sand mixture, 30–40 percent sand, and 20–40 percent peat. (Author's note: in place of the latter, we prefer to use peat-free potting substrate out of habitat-protection concerns). He does not mix in any soil for the long-term storage.

This is yet another process that provides support for the bacteria cycle theory: lactic acid fermentation and storing the greater portion of the valuable nutrient materials for 2–6 months (comparable to grass silage or sauerkraut production) provide high-quality nutrition for the life in the soil.

Quod Erat Demonstrandum: Grass Mulch over Larger Surfaces

The organic gardener Stefan F. ventured to prove that the Åkerstedt method also works on a larger scale. His attempt was successful, as you can see on the left. He employed the grass mulch method continuously from 2005 to 2010, adapting it to local conditions. Several of his colleagues who he is in contact with are attempting similar experiments with the same astonishing results. In 2009, he referred me to a document from the University of Hanover in which calculations were performed that came to the conclusion that his grass clippings could not have any fertilizing effect.

The practical results prove the opposite. The soil clearly did not abide by the results of the theoretical calculations.

> **Always Keep in Mind When Doing Any Sort of Mulching:**
> Grass and other plants that have been grown with mineral fertilizer contain (and thus provide) different cytoplasm, different proteins, and different living material for the edaphon and plants compared with grass and weeds that have spent generations living in harmony with and feeding on natural edaphon. Ecological food is always better, whether it's for you and your family or for your edaphon!

PROPER APPLICATION OF MANURE AND SLURRY

Manure in large quantities, as it is used today in agriculture, is first processed by anaerobic microorganisms (i.e., microorganisms that do not require oxygen) and then must be adapted by aerobic microorganisms (i.e., microorganisms that do require oxygen) in order to maintain healthy materials cycles in the biosphere. The same goes for every other type of organic waste, as you will see in the following paragraphs.

Initially, manure has biological value. In accordance with the cycle of living material, natural manure from purely organic sources contains far more useful energy than manure from conventional agriculture. The possibility of passing on good or negative characteristics or compounds from one generation of growth to the next is indeed a fundamental feature of *truly organic* agriculture. Currently, organic agriculture does permit certain amounts of manure originating from conventional operations to be applied if the organic operation does not produce enough on its own. But the guiding principle of organic-biological agriculture is that life can only arise from life, and that can only take place through the closed cycle of living material—not by ceaselessly incorporating dissolved artificial fertilizer salts into the biological cycle.

Manure needs to be processed as effectively as possible. There are several possible ways to accomplish this.

While it is still as fresh as possible, before it begins to stink, you can apply it in a thin layer onto the soil (exposing it to the air) and bring it into contact with the soil by working it into its surface so that the edaphon can continue processing it as quickly as possible—before it dries out or is eroded—allowing it to multiply efficiently. Then the edaphon itself and its metabolic products will be available as nutrient sources for the plants.

If the manure must be stored, it should be stacked up and very tightly packed together so that it contains as little air as possible. Fermentation then sets in, which consumes whatever air is left, leaving the manure inactive afterward, which makes it possible to store the manure without significant losses (Preuschen 1991). You can also intentionally introduce air to encourage the microorganisms to multiply aerobically (e.g., with a small aquarium air pump in a stinging nettle–based hydrocultural system or with an agitator and aeration in a large slurry tank).

In each of these cases, you are avoiding the anaerobic, oxygen-poor decomposition processes that inevitably involve unpleasant odors and are also a sign of the presence of toxic materials.

Hot composting is also one of these methods. All anaerobic processes with the exception of fermentation are special cases in nature for situations where extreme amounts of organic matter accumulate with limited aeration. Thankfully, nature is able to deal with such situations—but it always means taking an indirect route through disagreeable and toxic intermediate stages that strain, restrict, decimate, or even completely prevent the aerobic, symbiotic digestive processes.

Manure is definitely not the only option, however, and not the best one either—as we have now seen and understood. This debunks the claim that raising livestock is a prerequisite for plant production because of the manure. This agricultural combination serves only to provide a use for the waste products of meat production; it does not represent the optimal method of plant nutrition.

WATER FROM RAINWATER TANKS

In a rainwater tank, water acquires the tepid warmth that plants love so much. Sunlight shines on and through the water while it is inside the tank. It becomes green. Algae and other microorganisms grow in it. Annie Francé-Harrar (1957) describes humus moisture as follows: "But it's not a matter of 'water,' it's a matter of a high-quality infusion rich in organisms and nutrients that the plant roots can optimally utilize" (9).

Rinsing your hands, gardening tools, and vegetable crops also brings new microorganisms into the tank and increases their numbers. This is made clear by the color of the water or its appearance under a microscope. Here is where the food is grown for the plants, fresh and full of life. They "eat" the plankton while it is raw and living. That is emphatically more than just tepid warmth and water-soluble salts!

In nettle slurry and all other infusoria cultures made of organic material, much more protein embedded in living structures grows than even in a rainwater tank. We just have to make the effort to learn when these microorganisms are at their high point in terms of vitality, from the point of their inception, and take into account the amount of sunlight, the temperature, and their supply of air and nutrients.

WINTERING THE EDAPHON IN YOUR GARDEN

After you collect, clean, and store the harvest in the fall, the plant beds empty out and accumulate large amounts of good, fresh "waste," which is often not much worse than what you have stored. It's best to think in terms of nutrients and quality rather than just in terms of manure and filth (although those can also go along with it).

When you're wanting to throw away waste—spoiled apples, potatoes, and carrots that you didn't eat and forgot in your cellar—you should remember that it's too good for the trash. The rule of thumb should indeed be that only the best is good enough—fresh, ecological, living, and full of energy, without artificial additives—for all living things, including the edaphon. Nevertheless, fruit and vegetable waste, lovingly cut up into centimeter-small pieces, is better employed as your best helper in your garden than thrown into the trash. But use only uncooked waste! I'll come back to this subject in more detail later.

In the fall, after the harvests, leave all weeds lying on the plant beds. You don't need to pull them out, either. Why should you remove all of the roots, all of the material that they have produced during a growing period, and all the material that they feed their special symbiotic helpers with from the soil right before the winter, when the edaphon also needs nutrients and energy over the winter so that it can give your plants the opportunity to grow healthily with renewed energy the next spring? After all, the green parts above the ground are only half of the plant. The other half sits beneath the ground, and both halves work closely together as a system to provide a supply of nutrients and water (see page xx). For us, the key is to produce organic "food" for the edaphon, and "waste" (material we can't make use of ourselves) serves this role particularly well if it is pure, grown without mineral fertilizers, fresh, and still living.

If it's available, feel free to add more weeds and spread your kitchen waste over it.

Then cover the surfaces extra well for the winter with hay, straw, or foliage. This provides food for the life in the soil during the dark season. During warm winters, almost all of it is eaten. If it freezes solid, the organisms withdraw to deeper soil layers or wait for warmer weather in a state of torpor, at which point they can awaken and multiply rapidly, as long as enough food is available. In the spring, you can rake the remains of this wintering layer to the side, leaving an excellent seed and plant bed available for immediate use.

Before the winter, I lay out up to 10 liters of fresh, raw fruit and vegetable waste per half square meter. I lightly sprinkle this layer with rock dust and then cover it with a layer of 2–5 centimeters of wood chips, foliage, straw, or hay. You should also still be cognizant of the nutrient value provided to the edaphon when using this covering material. Hay, even if it is moldy or old, contains more valuable nutrients for the edaphon than straw, which has already passed on its energy to the grains. Wood chips need to be processed by fungi first.

You'll need to modify your edaphon-feeding method in accordance with weather conditions. Continuous snow and a thick isolation layer made of loose, airy material greatly aid the edaphon in maintaining its life functions over the course of the winter. Here in southern Norway, directly on the warm sea, we usually have temperatures around the freezing point and almost no snow, so I have to provide a warm cover for my plots and additionally protect them with spruce branches from my Christmas tree. Poultry netting can also be effective because winter birds quickly figure out where the best winter worms are found.

You should also take dogs, cats, mice, and rats into consideration. Meat, fish, milk products, and leftovers from cooked food can lead to problems with these animals and must therefore not be used. The key is to provide the best possible living conditions for the edaphon over the winter. You can systematically spread fresh kitchen waste throughout the entire fall and winter. If there is a period of frost, freeze the waste outside in a bin, then spread it during the next thaw, covering it well as described previously. A household of two people can supply ½–1 square meter of garden area with edaphon food per week using this method (see page 69 of the color images).

SOILIZATION: PRODUCING YOUR OWN HEALTHY SOIL

The term "soilization" is derived from the Swedish *jordisering* (*jord* means "soil"). Börje Gustavsson developed an auger that lets you bore holes of 30–60 centimeters in depth and about 8–30 centimeters in diameter in your garden (around trees, for example):

> Fall and early spring are the most important times for a garden. This is when the supply depot for the plants growing in the garden must be filled under the ground. This takes place via "soilization," which refers to firmly

packing fresh kitchen and garden waste as well as natural fertilizer in bored holes 30 to 60 centimeters in depth around fruit trees and bushes, as well as in vegetable gardens, their borders, and greenhouses. Once the waste is broken down via decomposition and the plant nutrients within it are freed, the substance collapses in on itself and the holes can be refilled with fresh waste. [. . .] The holes are also well suited for supplying water. [. . .] You can withdraw soil for flowerpots out of older holes using the auger. [. . .] This is the best planting, farming, and recycling method in the world. [. . .] The auger provides optimal yields. (Gustavsson 2001, 2016)

This method, which you can think of as directly feeding the edaphon, can also be employed on the mulch or soil cover, a process described by Rusch as "surface composting." Rusch described composting using piles skeptically and practically as a waste of fertility in the wrong place; it should instead take place where the fertility is directly used, meaning around the plants. Rusch still associates the idea of composting with the mineral theory, according to which plants can only absorb light water-soluble salt ions. But if we presume that the life in the soil does not comply with the system imagined by Liebig but that the edaphon—as the vital, living "population of the soil"—effectively eats the fresh, ecologically healthy food and multiplies accordingly into living cytoplasm with all of its amazing contents, then we shouldn't be indifferent to whether salt ions or microorganisms are the cause of growth.

I am very much inclined to look toward fermentation here. Fresh, organic material that has been pressed together pretty firmly, with very limited access to oxygen, must lead to lactic acid fermentation, like in the case of sauerkraut. And using this method you do not plant in the "silage-soil," but rather next to it, just as Teruo Higa (2000) recommends with his *bokashi* method. It makes use of organic material fermented with effective microorganisms, so that the plant roots can decide when, how, and what they want to eat and process from it. Higa also ferments his kitchen waste. "A sweet-sour smell after one to two weeks indicates a successful ripening." He also buries the bokashi next to trees, bushes, and vegetables (Higa 2000).

Rusch did explicitly warn against burying organic material too deeply and with too little oxygen, because he quite rightly was worried about foul-smelling, toxic rot. But with some distance from the buried material and enough freedom of choice for the plant roots, it seems that the plant knows how to overcome this drawback. In any case, Rusch did not factor the process of fermentation into his model.

In all probability, it is once again our reductionist way of thinking and experimenting that causes these three processes to be delineated and separated from each other. Down in the soil, they probably take place at essentially the same time, if not symbiotically (Higa 2000, 25: coexistence of anaerobic and aerobic bacteria).

Soilization within a Container

If you need good potting soil or soil for transferring vegetable plants from one pot to another and already need it by early spring, you can (or must) begin the process during the preceding fall, either using the Åkerstedt method described on page 163, or else with the approach described here. This approach consists of soilizing the kitchen waste discussed earlier in bins in your cellar or carrying out the same process on a larger scale.

In the cellar, thoroughly mix together chopped up kitchen waste with garden soil and store this mixture in containers with holes bored in them for drainage and aeration. Cover the surface with a loose layer of moss or leaves. If the containers are stored under warm conditions, you can check whether anything has sprouted after only two weeks. If the cellar is cold, you will have to wait at least thirty days. Waiting longer doesn't seem to cause any issues either. Then pour the contents from one container into another to loosen them.

This method provides you with excellent soil. In a 3-liter container, you can potentially grow one or two tomato inflorescences in this soil. With larger containers of 6–10 liters, you can grow a decent amount of pickles. You should also water it with nettle fertilizer (1:10 to 1:5 diluted) and place a layer of fresh herbs and grass on top of it.

On a very small scale, this can even be done in the kitchen. The best way to do so is to take fresh (i.e., raw) leftovers from your ecologically friendly lunch, making sure that it doesn't contain any meat or fish remains (peels from organic citrus fruits and coffee grounds can be included). Chop these up in a food processor for a couple of seconds along with plenty of water until they are coarsely mashed together. Then strain out the water and let the mash fully dry out. Once the coarse material is relatively dry, mix it with an equal volume of living, good garden or compost soil. This also needs to be fairly dry, but it mustn't be dried out completely. Mix the dried mash and compost soil together in small quantities (about 1–2 liters) in a 10-liter container, until no clumps remain. Clumps can only be processed by anaerobic microorganisms due to the lack of oxygen inside them, and this causes unpleasant smells (such smells are a sign that the soilization process is not proceeding correctly). You can also strain the mixture through a 5-millimeter sieve or spread it out in the cellar or in a shaded area outside in order to dry it. You can add small amounts of limestone, wood ash, or stone meal if you wish.

Always remember to keep the living organisms that you're working with in mind!

In my experience, surprisingly enough, you can move all of your seed plants to this life-filled soil almost immediately after doing the mixing or, at worst, after a few days or weeks pass, giving the microbial community time to consolidate. You can also cover the freshly mixed soil with a few centimeters of mild sowing soil and plant your seeds in it so that the young roots reach the nutrient-rich edaphon somewhat later in their growing process.

How long this kind of soil can be stored without losing some of its vitality remains to be discovered by competent microbiologists working on humusphere-related subjects.

> **Why Aren't Researchers Taking This Up?**
> There is a lot of room here for individuals to modify, further develop, and, when necessary, correct these suggestions based on their own experiences. And it's an interesting potential new field of research for agrobiologists and plant physiologists. Our tax dollars would be far better spent on research along these lines than on genetic experiments with their incalculable risks that even biologists are not able to estimate.

MY EXPERIENCES

The following is a description of my own experiments and methods of growing vegetables on the basis of optimally caring for the life in the soil and the edaphon. I have been carrying out some of them for decades with persistent success. But since garden and climate conditions can vary, as can the types of living material used and the edaphon in the soil, I recommend starting with a small test (using garden cress perhaps).

Reusing Fruit and Vegetable Waste

When I first started my involvement with permaculture, I began with small-scale experiments in my rock garden in southern Norway. First, I sorted my organic waste from the garden, the surrounding heather-filled areas, and my kitchen into pure "ecological" remains (relying on store-bought items as infrequently as possible) on the one hand; then I sorted waste originating from "normal" store-bought things (i.e., conventional food) on the other. The former results in pure, clean compost or soil cover, while the latter produces contaminated organic waste. All of my experiments described here solely used the clean organic remains (originating from organic food), and I made sure that they hadn't yet spoiled (i.e., begun the process of rotting).

My first experiment: I laid a wooden frame on a wide stone wall that had been built with stones from the nearby scree field generations earlier. This stone wall now separated my stone garden from willow that had been cleared of rocks by hand and by using horses. Gradually, I filled the wooden frame with organic material from my kitchen and the rest of my property to a height of about 20 centimeters and placed kale plants with root balls inside. In fall and all the way into the spring, I had excellent kale. More than twenty years later, potatoes, blackcurrants, and Jerusalem artichokes are still thriving on top of all these stones.

Soil formation in nature with lichens and mosses happens somewhat slower. But the principle remains the same.

After many years of such experiments, as well as others, I had developed the following approach, which I am happy to pass along.

Highly Efficient Reuse of Waste

If you have some time while preparing meals, you can develop the habit of also thinking of your best helpers in the garden and cutting up your waste into smaller pieces. (It's best to bring the waste to the garden directly from the kitchen.) Just like we do, the edaphon prefers food that is as fresh as possible. It should also be clarified that there is hardly any organic waste anymore that is still of a high enough quality to serve as healthy food for the edaphon. Decades of dispersing chemicals and transgenic elements has contaminated all of our foodstuffs to a greater or lesser degree over the decades. So be careful to only apply fruit and vegetable remains that are of the highest possible quality (ideally organic or from your own garden).

For two people, I collected almost a week's worth of plant remains (of organic quality if possible). That provides 7 or 8 liters of remains, which should not be packed densely and moistly enough that they start to smell.

I then spread the remains over ½–1 square meter of my garden. To ensure that the neighbors don't start to wonder, to make the area more aesthetically pleasing, and—most important of all—to protect the edaphon, I then cover the surface with straw, hay, or grass. You can even put down a rather high pile of weeds, cover it on four sides with wet newspaper, distribute the kitchen waste through it, and cover it. You can then tear a hole in the newspaper and place a zucchini, pumpkin, strawberry, or cabbage plant in it along with its root ball. Potato plants are also an option here.

The whole thing should sit peacefully at first. The important thing is that the setup doesn't dry out, because moisture is essential to the edaphon and because the dry newspaper would cut off the upper layer from the lower layer, stopping any exchange of material and organisms before conditions can get moist again.

You can make your own plot of this type by bringing out ½–1 square meter of fresh kitchen waste each week and carefully covering it. Depending on the season, it can either be immediately planted or left out for use at a later point (see the top and bottom of page xx of the color images).

You can try this method directly on any sort of grass surface, on any surface that is growing weeds, and—with a little more experience and organizational material—even on sand, stone, or concrete. Asphalt should be avoided, though, or covered with plastic if absolutely necessary. Just keep in mind that both asphalt and plastic are toxic!

Remember that any patch of ground that is not covered with green plants and their roots will be unable to store solar energy. Every single weed works like a solar cell. They collect the energy from the sunlight (even if the weather is overcast) and pass it on to the plant roots, fueling all of the edaphon with renewed metabolic energy.

Let's do a quick calculation: 7 liters of waste per week times 52 weeks in a year over 1 square meter of garden soil equals 52 square meters, enough to provide excellent garden soil and harvests that would probably be impossible otherwise.

My Onion Plot

This is how I plant onions without any sort of soil-digging work.

I cover an approximately 1½ x 2-meter area with remains and waste from jarring, pickling, and freezing my fruit harvest to a depth of about 1–3 centimeters. On page 69 of the color images, you can see pomace from elderberry juice in the middle of the plot on the upper left, potato skins on the upper right, and regular kitchen remains no more than a week old underneath, sprinkled with stone meal. Similar material is covered with 1–2 centimeters of mulch made from chopped twigs and kept in storage for half a year to the left and right and finally covered on the left with cut-up spruce twigs (or chicken netting), because birds are well aware of where earthworms can be found.

In the spring, the cut-up twigs are removed. Everything that was previously underneath will have disappeared, even after the mildest of winters. If the feed was too coarse and the winter was too cold, the materials' removal happens more slowly, but the edaphon work and eat during any given fall. During frost, however, it retreats further back into the humusphere or encysts itself to "hibernate." But it always resumes its activity in the spring. These biological processes will need to be adjusted based on experimentation for a given climate zone and for any specific given garden or field. If the winters are too cold, you will have to insulate the waste with dry material, such as hay. Snow also makes a good winter cover.

Assuming it worked properly, there will be nothing left after the covering except for fine, crumbly soil: everything has been "eaten" by the invisible edaphon. If there are a few bulky chunks of remains left over, rake them until they break down. I immediately cover the fine, crumbly, weed-free soil with an old felt cover temporarily so the edaphon isn't exposed to extreme sunlight or heavy rainfall.

I then work the soil only once with a soil loosening pick and place the prepared, needle-thin onion germ buds inside. I prepare the Ailsa Craig onions, in the cellar late in the winter. I place the young plants at 10-centimeter intervals with 15 centimeters of space between rows. Between the rows, I loosely place 10 centimeters of freshly cut grass and herbs from a meadow. This grass cover keeps nearly all weeds from sprouting underneath.

Two months later, my onion shoots are as thick as a pencil and the original grass I placed between the rows has almost disappeared. I then place another 10 centimeters of fresh grass between the rows, making sure not to cover the shoots. Nothing else needs to be done until fall. After the harvest, I cover the area once again for the winter with a layer of fresh kitchen or plant waste, leaves, hay, or straw that is 2–3 centimeters deep and cover it with fir twigs or a bird net.

This method has provided me with about 18 kilograms of onions (or, alternatively, 14 kilograms of leeks or 11 kilograms of celery) for seven consecutive years from the same plot of land, without any signs of the soil becoming sick or weak.

I interpret this to mean that repeatedly intensively feeding the edaphon maintains healthy fertility in the soil.

My Kitchen-Scrap Porridge

To make kitchen-scrap porridge, I fill a blender three-quarters full with fresh organic plant material from my kitchen, garden, and from nature (much like with soilization, described on page 170). I fill it the rest of the way with rain or pond water—not tap water if at all possible—and let the blender run on a high setting for a couple of seconds, until the mixture forms the consistency of porridge. Then I strain the water and use it, diluted to 1:5 or 1:10, to water my flowers or feed the edaphon in my garden.

I then mix the porridge, which should be as fresh and dry as possible, in equal parts with semidry garden soil or compost in a large bucket. It's important to make sure no clumps form during this step because clumps lack oxygen, which is counterproductive for the soil life. You can avoid clump formation by first pouring the vegetable porridge into a towel and wringing it out and, if necessary, sifting the mixture through a large sieve (see the bottom of page 185 of the color images).

Chlorophyll Water

To make chlorophyll water, I blend a handful of grass and herbs together with 1 liter of water in a blender for a few seconds, then strain the green water and use it to feed tomatoes, beans, and cucumbers until they are ripe (see the bottom of page 71 of the color images).

In my experience, houseplants react almost immediately to being watered with this sort of chlorophyll water. Some already dis-

A look into the author's experimental garden. All of these plants were grown in sand with a neutral moss cover and exclusively supplied with chlorophyll water.

This was also practically tested by the author: floating plants in algae and chlorophyll water.

Lettuce growth in expanded clay and chlorophyll water.

played a visibly deepening green coloration in their leaves the very next day. Others, which had previously sat inactive or even hung in the window in a sickly state, began to grow vigorously. Even my experiments in which I planted vegetable saplings with their small, root-filled soil clumps in pure, nearly sterile substrates (such as sand) and then fed them chlorophyll water exclusively yielded surprisingly positive results (see pages 71 and 179 of the color images).

Bringing the Forest Back to the Fields: Sawdust and Bark in Agriculture and Gardening

I have also experimented successfully with wood waste, such as sawdust and bark (see the bottom of page 69 of the color images). Sawdust is an excellent humus carrier, and bark works nicely as well. But neither should be used when fresh; they first need to undergo a composting or aging process that can take anywhere from weeks to years that dampens and restructures them.

In my experience, slurry and sawdust mix together and compost very well. After it is allowed to age, I apply it in the fall as a mulch layer with a depth of up to 10 centimeters and rake away the largest chunks in the spring. This is the best protective and nutrient-providing cover for the life in the soil that you could possibly hope for.

Of course, it's important that the sawdust only come from sawmills and from untreated wood. Sawdust from cabinetmakers' shops mostly originates from treated wood (e.g., particle board) and is nothing but hazardous waste. Bark compost makes an excellent fertilizer, but fresh bark has herbicidal properties and greatly inhibits seed germination (which is why fresh, cut-up bark is also used as bark mulch under bushes and beds of perennials). Neither sawdust nor bark should be buried in the soil; this will cause peat to form due to the lack of air.

Further Experiments

In other experiments, I have fed my plants with microorganisms in all kinds of different ways: with aerobic infusoria cultures from stinging nettles that never smell unpleasant and with every imaginable mixture of fresh green material and organic fruit material with water. My plants grow in solid manure silage, grass silage, or in our sauerkraut (even with its own juice) mixed with living soil.

Lettuce Grown Unconventionally

I have also experimented with lettuce, grown in a mixture of sauerkraut and garden soil: one part sauerkraut with juice mixed with two parts finished, but fresh, living compost (see the top of page 71 of the color images). The result: the lettuce plants grew vigorously and well. You can achieve similarly good results with farm silage.

Out of the Kitchen, Into the Garden

I carried out the following experiment: I blended ½ liter of kitchen waste together with 1 liter of water in a blender and spread this mixture over a square meter of dry moss once per month for a year, starting in October. This produced 6 liters of kitchen waste per square meter per year as food for—I assumed—the edaphon living underneath the moss. The result was thick green grass growing in the area once again.

Let's calculate: a family of four produces about 0.3–0.6 kilograms of kitchen waste per day (not counting fish or meat). This waste has the nutritional value of approximately 3–6 potatoes. This allows 100–200 kilograms of organic material to supply the edaphon living in about 15–30 square meters of garden soil, green area, or flower or vegetable beds per year—and it will provide an especially good harvest and a convenient way to get rid of your kitchen waste. It's a good idea to cover the "fed" area of the vegetable garden with a loose layer of grass to protect the life in the soil and its food from drying out.

Stinging Nettle Liquid

Many of us surely remember the "hay infusion" experiment from elementary school or later biology classes. You place some hay or grass into a bottle with water in it and leave it on a window sill for three days. This gives rise to a so-called infusoria culture. Old water from flower parts or stinging nettle liquid also make great nutrients for your garden.

However, the latter ordinarily contains so much water-soluble nitrogen and potassium that you have to be careful not to use too much and speed up growth excessively. Stinging nettle liquid smells horrible and works too well if you use it in excessively high concentrations. The standard explanation for this in books on ecologically oriented gardening is that it is providing too many NPK nutrients. I believed this myself for a long time and didn't use any stinging nettle slurry at all. I used only stinging nettle liquid in small quantities. After all, I didn't want to overfeed my plants with water-soluble salt ions! But finally, I realized that there was a relationship between hay infusions and stinging nettle water and I examined a drop of the latter under a microscope. What I saw were thousands of single-celled organisms, amoeba, and flagellates moving around. And remember that we can only see the millions of even smaller bacteria under a magnification of at least a thousand times, if at all, and even then still just as tiny points moving rhythmically together. Those aren't water-soluble salt ions, they're living creatures—millions, billions of living creatures!

This kind of infusion serves as a "living protein soup" for a few days or weeks—as long as there is still air and nutrients available—that the plant roots "eagerly slurp down," as Hamaker put it in Tompkins and Bird's *Die Geheimnisse der guten Erde* (*Secrets of the Soil*; 1998). The first sign that the oxygen is fully depleted will be it starting to "typically" stink, as other metabolic processes will be starting up that make use of anaerobic microorganisms (i.e., microorganisms that do not need oxygen). These processes tend to produce toxins, much like the untreated and unaerated liquid manure that is frequently used in agriculture (see the discussion starting on page 75).

Since I looked through that microscope, I've also used stinging nettle liquid. I no longer fear the possible dissolved salt ions, as they are far from being its primary component, if indeed they play any role at all. A small aquarium pump placed in the stinging nettle container is enough to turn the "in-house poison factory" that Rusch extensively warned against into an excellent, rich infusoria culture with large amounts of living cytoplasm that the plant roots can absorb. This isn't a case of mineralization, rotting, or composting; in fact, there's no breakdown of any sort—just an aerobic, living transfer of organic cell contents to living microorganisms, or, as I call it, syntrophic living conversion. And it doesn't smell bad at all!

EFFECTIVE MICROORGANISMS AS FINISHED PREPARATIONS

All of the methods and experiments described in the previous chapter are based on the principle of fostering and keeping healthy the naturally occurring microorganisms in the soil and in the plant. But if you don't have time for all that, you can still promote the life in your garden with finished preparations.

A relatively large variety of so-called soil improvers, plant fortifiers, and compost starters with effective microorganisms are commercially available. But what I have never found, with a single exception, is an indication that plants can absorb and process the microorganisms in these products as nutrients in living form (as long as they are still alive). "A systematic application of starter cultures made from effective microorganisms to the soil and plants in your gardens, orchards, and fields reportedly causes practically all cultivated plants not only to grow more healthily, but also makes it possible to store them for a longer time. Combined with sufficiently large quantities of organic material, which accrues in large amounts in gardening and domestic chores as well as in farming, the MicroVeda® effective microorganisms speed up humus formation in the soil and stabilize it in a sustainable form. Plants primarily feed themselves through their roots from living single-celled organisms, soil fungi, and soil bacteria. This process is known as 'endocytosis' in science" (Rateaver and Rateaver 1993; Rateaver and Rateaver 1994; see also www.mikroveda.eu).

Series of experiments carried out by the author. Left: Soilization in a container. Top: Germination test with cress. Right: Tomatoes and pumpkins growing in soilized substrate.

Left: Producing "porridge" out of plant material and kitchen waste mixed with regular soil at a 1:1 ratio produces an excellent fertilizer. The strained water can be used for watering plants.

Section 4

Final Thoughts and a Look Toward the Future

— CHAPTER TEN —

My Vision

THE COURAGE TO EMBRACE NEW WAYS OF THINKING AND ACTING

Attempts to offer practical advice for agriculture in general or even just for small-scale gardening often fall into the old habit of turning to the artificial fertilizer model. But I believe that the alternatives to that habit in fact represent our real hope of reaching an understanding of the cycle of living microorganisms. Allow me to once more summarize my views and thoughts.

We come up with models to help us understand the world: theories, worldviews, scientific approaches, religions, and so on. These models lead us to conjectures and then to experiments whose goal is to confirm, expand, and "prove" them. With the help of these models, we construct an apparent reality, one we hold to be true as we come to see it more and more as the only possible, real, concrete explanation for things, while we view other models as false (even if they were originally on equal footing with our chosen model when the problem or subject was first being considered).

Eventually, one model catches on more than the others, gains support from more people, and sooner or later displaces the others in the arena of what we see as truth. As time goes on, these "majority models" cease to even be examined. No one questions where they came from or how they were developed or what their current state and future look like. And eventually, the mistakes that any given explanatory model inevitably induces cease to be recognized as mistakes at all. Instead, they simply continue to be pursued within the framework of the one model that is assumed to be correct, both in everyday life and the academic world. Dangers brought about by these mistakes are rarely recognized as such, and even when they are, they are downplayed or (seemingly) refuted for as long as possible. Particularly when the commercial interests of certain groups are threatened, attempts are made to pay for research or publish academic books and "experimental results" that prop up the "one true" doctrine for as long as possible.

One such model is the mineral theory. It has now dictated the standard approach for a hundred years, and during that time it has poisoned the entire biosphere with inorganic, synthetic material. Many of the catastrophic effects of these chemicals are well known while others still wait to be discovered. The model has progressed under the direction of biochemists to the point where even living substances and their ability to organize themselves are being manipulated, which is sure to lead to biological catastrophes that we haven't even begun to consider yet.

The way that we treat our soil, in accordance with the mineral theory model, has devastating effects on the biosphere, effects which have long been impossible to ignore. We have uncritically followed an extremely simplified and reduced model of plant nutrition and made it the basis of our entire chemo-technical agricultural system—and in the process we have poisoned our food, ourselves, and the entire biosphere.

This book has presented possibilities and methods that could allow us to avoid these dangers. In order to free ourselves from the nearly fatal embrace of the technological worldview, we will have to drastically change the way we live. Whether that will prove to be possible and what sacrifices will be demanded from us depend on each and every one of us and what we want to achieve. Here is my suggestion:

We must develop practices that promote frugality, welfare, and cooperation and view these things as the benchmarks of success. Competitive behavior, unchecked growth, heavy resource consumption, and universal employment solely based around a desire for money are practices that need to be dismantled. Every job in the modern workforce chips away at a little piece of our biosphere, as these jobs, and just about everything else we do, are built on the basis of a model that encourages greed and competitive thinking. The fact that earning money is necessary to feed yourself and to survive gives this system unlimited power over every aspect of life—and at the same time, it's wearing out the biosphere to the point of its destruction.

There have been many attempts to change this unhealthy system. But for the very reason that these attempts propose systemic changes, they are seen as hostile threats to the prevailing order, and treated by modern society accordingly. Furthermore, all of these individual attempts need to be pooled and adjusted based on each other so that a consensus can be reached, something that takes time. Unfortunately, technological development and global wealth redistribution has thus far been a faster process.

To me, one thing is certain: the claim that without artificial fertilizers, biocides, and genetic modification, the world will starve is a pure and purposeful lie.

In 1975, Professor Dörner from the University of Giessen demonstrated how a one-sided scientific education can inhibit people's comprehension of complex and unexpected interrelationships. Students with significantly above-average intelligence were unable to adjust their preconceptions even in the face of changing circumstances. No matter what happened, they stopped gathering

> **We Have to Decide**
>
> Do we want a globalized society that is in competition with the environment, a competition it will ultimately lose? Or do we want a society that can sustain itself, peacefully and healthily, in cooperation with nature?
>
> The visionary ideas in our heads today can be the guideposts showing us and our children the way toward a future reality tomorrow!

information and were unwilling to correct their preexisting ideas (von Haller and von Haller 1976).

Forty years later, as we run the risk of drowning in a veritable flood of information, we have barely begun to address this problem on any sort of large scale. The field of agriculture—even the modern incarnation of ecological agriculture, I would intentionally add provocatively—is not yet ready to take on this challenge.

At this point I would like to thank Per Krusche (1982) and all of his collaborators for the fantastic book *Ökologisches Bauen* (*Ecological Growing*), which inspired me in numerous ways. I still consider this book to be one of the most comprehensive and thorough guidelines for the ecological movement available, with truly inspirational drawings, concise data, and an excellent bibliography. I am equally grateful to the deceased Wolfgang von Haller, who sent the entire catalog of his publishing house, *Boden und Gesundheit* (*Soil and Health*), to me in Norway. Along with the writings of Hans Peter Rusch and Hugo Schanderl, these researchers gave me a fascinating glimpse into the biological thinking of the twentieth century, which continued to be pursued by newer luminaries such as Bill Mollison, Margrit Kennedy, Lynn Margulis, James Lovelock, Teruo Higa, Sepp Holzer, and many others.

We need to acknowledge the third stage of living material as defined by Rusch (see page 122) as something that exists and is part of the cycle of living material. Until the cycle of living material manages to take up its place in biology and nutrition science as a fundamental, scientifically proven explanatory model, our food will continue to be chemo-technically based and thus inadequate.

To me, it seems like the future of our children depends more on the courage of biologists than the audacity of technologists. We finally need to change our out-of-control, societally and ecologically damaging economic system. And we need to be willing to personally play a role in this paradigm shift, even if doing so can often be painful.

— CHAPTER ELEVEN —

Concrete Steps

OUR AGRICULTURAL METHODS NEED TO BE REFORMED FROM THE GROUND UP

I must concede that it isn't easy to liberate oneself from the model of water-soluble plant nutrition—for understandable reasons. Most significantly, if your entire education and planning decisions have been based on this model, it takes great effort, lots of courage, and an independent way of thinking and critically questioning to free yourself from it and give consideration to a new one. But perhaps it's easier when framed like this: A radical new understanding of the fundamentals of biology in relation to both our agricultural practices and our food means *recognizing the importance of plants, animals, and humans working, functioning, and living together, and creating conditions that allow it to happen.*

Our technologically based agricultural system rules out many sensible possibilities. It has become so complex (a consequence of all the failure and neglect when it comes to biological processes) that only highly qualified specialists are able to make it function (and bring about the eventual catastrophe). Farmers, and even politicians and consumers, have been effectively incapacitated and are no longer competent to really assess and judge agricultural subjects. Ultimately, they are all just stooges for lobbyist groups and the "free" market.

What I hoped to make comprehensible in this book is that any biologically based system needs to be focused in one way or another on edaphon. One hundred seventy years of endless experimentation has shown beyond doubt that applying Justus von Liebig's mineral theory—which has long been obsolete, if not outright false, due to how narrowly it defines things—to biology and plant physiology is not the way to success. Whether or not it's intentional, conscious or unconscious, it inevitably leads (first in theory and then in practice) to attacks on living processes. In my view, we could just as easily describe the natural sciences as the "dead material sciences," since they deal with the living processes coordinating all of the systems in the biosphere to a very limited degree—especially agriculture.

Far more dangerous, however, is the widely held but false notion that science can guide and direct these living processes via technological and chemical means in a way that makes them more sustainable or forward-looking, or just "better." To quote Jürgen Dahl, this idea should be ascribed to "the audacity of the clueless" (Dahl 1989).

HUMUS SAPIENS: THE SOIL KNOWS WHAT IT HAS TO DO

The grave dangers posed to our food supply and the biosphere by technologically based agriculture make it absolutely imperative that we replace it with a superior, biologically based system. "Humus sapiens" (or, "the wise soil") is my working title for this effort. Here is a proposal for how we can initiate this challenging task among students and apprentices of agriculture: I propose avoiding an either-or outlook when it comes to technological measures and biological methods, instead striving for a both-and outlook.

All of the "biological challenges" mentioned thus far are surmountable, and many of the successes have been documented in practice, from the very beginnings of human agricultural activity all the way through today's most recent experiments. What is not achievable any longer is complete freedom from chemicals and toxins, because practically no organic material nowadays, living or dead, is free from anthropogenic chemical contamination. Our quality scale should therefore be oriented around the highest purity that is in fact possible. In the end, no amount of technological "improvement" can justify the usage of mechanical or chemical "aids." *There is no better and more healthful source of nutrients than fresh material provided to pure humus.*

All claims to the contrary are delusions based on science that has long ceased to be independent and that has blind spots, on our countless technologies that are hostile to life, and on a boundlessly avaricious money-based economy. We "design" our food today to meet the needs of the "free" market. If we want to even get close to food that can be called organic and healthy again, we will have to change the structure of our society—or start doing our gardening ourselves.

Our Future Will Be Decided by the Earth's Humus Layer

An intact humus layer is the only possible basis for healthy, toxin-free agriculture. It is the result of the functions of 1–2 metric tons of edaphon per 1,000 square meters of field soil.

So what concrete steps are necessary to propel us forward in a positive direction? I suggest the following four steps.

1. Unite All Theories and Models

We can explain the metabolic processes within field soil by drawing on three very well-defined theories and contrasting them with each other:

1. Justus von Liebig's mineral theory, which is the result of the direct application of the principles of chemistry to agriculture and plant physiology. I will refer to it here as *techo*.
2. Rudolf Steiner's biodynamic theory, whose goal is to understand the objects and phenomena that make up our world holistically. For this theory, I'll use the abbreviation *biodyn*.
3. Hans Peter Rusch's organic-biological theory, known among doctors and microbiologists who work with the cycle of living material as *orbio*.

We should compare these three models with actual conditions in the humusphere and the edaphon. Doing so will allow us to recognize inauspicious and incorrect measures we might be taking and, when appropriate, correct them. But each model can maintain its own quirks and research focuses. Models like these make it possible for researchers and students to bring order to the chaos of information with their model of choice, but before doing any research that is reliant on a particular model, it is essential to demonstrate that the model in question is in fact appropriate for the task at hand. The constant pseudoscientific debates over "ultimate" truth, over still-uncertain scientific evidence, can thus be contained or avoided entirely, allowing the energy to be applied elsewhere. *Consolidating* agricultural theories with the goal of maximizing the edaphon seems to me like a very promising approach.

2. Designate New Areas of Research Emphasis

a). In Practical Agricultural Research

If we make plant endocytosis the basis of our future conception of agriculture, we'll be able to move toward well-defined explanations and finally clearly and firmly sever ourselves from the purely chemical model and its devastating practices.

As described earlier, it's very possible to raise and multiply your own very good edaphon in your garden and to reach the sorts of yields mentioned throughout this book. The way to create the best possible biological conditions for microorganism multiplication on a large scale—over big agricultural areas—has been known and put into practice for years now by the producers of preparations that use effective microorganisms.

More edaphon means more soil structure, more protection against erosion, and more water-retention capacity. It also allows for increased plant and root growth, more efficient utilization of solar energy through the chlorophyll, and a greater supply of energy all the way down to the roots—this in turn leads to more edaphon, more food for the plants, more soil structure, and in the end (at least according to my vision) to decidedly higher yields than just 60 percent since 1950. And it would manage all this with significantly lower energy costs.

The use of organic material in agriculture can be microbiologically described and researched with clear, practical suggestions for how to carry it out based on the principle of maintaining and strengthening the cycle of living material. Any use of chemical and technological "aids" must, above all else and without compromise, avoid any harm to living material. We must avoid any reduction in their diversity and thus in the variety of their functions, and we must make efforts to ensure that the edaphon multiplies. This not only has the aforementioned potential to greatly increase yields but can also help with energy conservation, something that is much discussed but has not yet been realized anywhere. You can find good examples of this in the great cultures of the past, in the family of gardeners described in this book, and in Siegfried Lange's winter rye. All that the small-scale experiments in this book are waiting for is systematic research and large-scale implementation!

b). In Microbiology

The exclusively physical- and chemical-oriented model of plant nutrition is so firmly rooted in our minds that new approaches in either research or in practice have long been neglected. Microbiology based on the humusphere, on the other hand, offers unexplored and, it seems to me, entirely unimagined prospects. We must therefore be ready to educate ourselves about the unimaginable power of the smallest forms of life until we have understood them. Then we can start to live together *with* them in the biosphere and reach a state of true natural balance that will ensure the continued existence of our biosphere for future generations. The use of artificial chemicals and energy in agriculture is damaging to the biological processes of plant nutrition that we have only recently become aware of, leading to a reduction in our harvests and our energy efficiency. We need to realize that a one-sided view of certain technologies can be dangerous, as this approach has proven again and again to lead to constant conflict with the most essential aspects of our own living conditions, which were once biological in nature. I believe that this is what so often leads to major problems nowadays, and may ultimately lead to a global biological catastrophe.

It is possible to use our technologies for the greater good rather than to the detriment of our planet's entire biosphere only if we do so on the basis of broadly encompassing biological knowledge and a resulting deep biological consciousness.

Reusing organic material is a very old concept, but nowadays it has become a nearly insurmountable problem. Thanks to the complete contamination of all organic materials with up to a hundred thousand artificial, synthetically produced, or inorganic substances, any recycling of organic material simply causes these substances to recirculate and accumulate, in addition to allowing low-quality proteins and other low-grade nutrient materials to continue circulating.

My hope is that if agricultural microbiologists work with plankton and edaphon, they will be able to offer relatively pure microorganism cultures that function best as living cell and cytoplasm cultures for the multiplication of edaphon and as direct nutrients for plants. The literature contains countless examples that point in this direction, such as the algae spirulina and chlorella, which are bred both in nature and in cultures; *Methylococcus capsulatus*, a microorganism that helps produce large quantities of organic proteins from methane gas, oxygen, ammonia, and salts; effective microorganisms; or the perfectly ordinary duckweed, which only requires a tenth as much cultivation area as soy in order to provide the same amount of protein.

Of course, none of these production processes of biotechnological protein and living microorganisms solve the aforementioned issues of contamination or decreased biological quality, but they may be able to steer things in a positive direction with the help of the cycle of living material model. The development of effective microorganisms seems quite credible to me because it was based on solid microbial science.

c). In Plant, Animal, and Human Physiology

Like the model used in plant physiology, the explanatory model of human physiology also needs to move away from one-sided reliance on the chemically oriented way of thinking. Plant and animal physiology have much more in common and are much more closely interwoven than what is currently taught in our schools and colleges. The Gaia theory and the endosymbiotic theory of the twenty-first century allow us to build and expand on the conviction behind Müller and Rusch's organic-biological model: "Only life can create life" (Müller 1983).

3. Reform the Education System

My view is that all of these works should find a place not only in the college lesson plans of agricultural scientists and plant physiology researchers but also in school curricula and biology classes. Hans Müller's and Hans Peter Rusch's concept that only life can create life, for example, would fit well into a school lesson. You can also easily reproduce these experiments on your windowsill using simple methods. If you are able to practically comprehend what living substances that have been grown in a good environment are, you will begin to appreciate their value, and this finally opens you up to the realization that our purely chemical-technological methods ultimately mean the poisoning of the life in our soil and the destruction of organic material of all types. The best way to reach that point is to study the works of Rusch, Schanderl, and Margulis while also carrying out your own experiments in your garden at home or school. The examples given in the practical section would almost all be well suited to that sort of thing.

My research library for progressive teachers and students is as follows: The books of Hans Peter Rusch, such as *Bodenfruchtbarkeit—Eine Studie biologischen Denkens* (*Soil Fertility—A Study of Biological Thought*; 1968, 2004) and *Naturwissenschaft von Morgen* (*Natural Sciences of Tomorrow*; 1955). His thirty or so journal articles from the period between 1949 and 1974 (most of which appeared in the journal *Kultur und Politik* [*Culture and Politics*]), along with Annie Francé-Harrar's *Die letzte Chance* (*The Last Chance*; 1950, 2007) and *Humus—Bodenleben und Bodenfruchtbarkeit* (*Humus—Soil Life and Soil Fertility*; 1957), comprise a widely underappreciated research library. They provide an excellent framework to any progressive agricultural or gardening student or professional with a serious interest in the development of James Lovelock and Lynn Margulis's Gaia theory and in sustainable farming and gardening in general.

4. Establish New Quality Measures for Foodstuffs and Agricultural Products

It would also be important to put our impression of what constitutes good and healthy food to the test. According to the knowledge presented in this book, any food on or in field soil—anything that we eat ourselves, that we feed to our pets, or that we use as "fertilizer" for plants—must be judged on the basis of the living energy that it still contains and can pass along. This value can also be expressed monetarily—and raised considerably. This would give us a method that would allow us to apply quality controls in terms of nutrient and "vitality" content to all kinds of "refinement processes" in food technology, everything from preservation methods to biotechnological procedures. And any normal citizen could carry out these tests, making him independent of the food production industry.

The Path Isn't New, It's Just Buried

This is how we can find the path back to what were once natural, organic methods, methods that can and must be built on and confirmed with our newest microbiological discoveries. Rusch has already prepared the path. The fact that we haven't more diligently made use of it, keeping it active and open, has led it to become "overgrown" and invisible, while traditional (and sadly, to a certain degree at least, organic) agriculture has become more and more firmly ruled by the chemical industry.

My goal in this book has been to find that path again and to demonstrate it to you, dear reader. I also wanted to show that this path is still navigable and leads onward, just as Rusch hoped, to a "science of tomorrow," to a self-sufficient form of organic agriculture with the cycle of living material as a fundamental building block.

If the ideas collected here reflect nature and its reality better than the mineral theory does, then our goal needs to be what Annie Francé-Harrar offered us in her description of ideal soil with an edaphon content of 20 percent (see pages 53 and 61). With 1 percent residual edaphon, no amount of artificial chemistry, genetic engineering, or anaerobic toxic manure will do any good.

Incidentally—if you will humor me a brief philosophical digression—the cycle of living material has a whole other aspect to it: our fear of death could be soothed by the idea that every living thing carries with it the task of keeping life as healthy as possible and that in the end, it passes this life on to others in the form of its smallest living components, allowing them to live new lives of their own.

This also helps keep all forms of life from multiplying excessively and thus leads to a healthy balance in the biosphere. It wasn't until humanity came along with our industrial method that this balance began to change, a change which has since continued to an extent that poses a danger to our entire planet.

THE CYCLE OF LIVING MATERIAL AS THE BASIS FOR ALL ORGANIC THOUGHT AND ACTIVITY

It's not a matter of my being "right" with my model. There can be no *single* explanation, no *single* model for all of science. The model I suggest here for further development does, however, make it possible to find an explanation for these many examples of higher yields per unit of area, something which the other model cannot account for. I also see incorrect explanations as the reason why these "unbelievable" facts aren't widely known and no one has yet gotten to the bottom of them. Without a model to explain them, we are unable to grasp facts, even if they are right in front of our eyes, so we perceive them as untrue or completely nonexistent. Explanatory models substantiate things for us, whether they be important, true, or false.

Let us return once more to Hans Peter Rusch, who was the first to call attention to the possibility of a cycle of living material. As a contemporary of a one-sided, technology-dependent, industrial society, he, like Hugo Schanderl and so many others, worked against the erroneous, purely reductionist dogmas of the established scientific community. His thoughts, theories, and practical experiments were dismissed and banned by that community, often in a discriminatory manner (see Vogt 2000), but they remain accessible in libraries. In the appendix, I have included a list of related literature and new editions when present. But of course, this is only a selection.

As for the renewed study of Rusch, let me note the following: Rusch's lectures from 1949 through 1954 are short and are all collected in *Naturwissenschaft von Morgen* (*Natural Sciences of Tomorrow*, 1955) for easy reading. His major work, *Bodenfruchtbarkeit: Eine Studie biologischen Denkens* (*Soil Fertility: A Study of Biological Thought*), which was released in its eighth revised edition in 2014, is a natural starting point for a deeper understanding. However, it's not entirely easy to understand what Rusch actually means by the term "living material" because the definition must be seen in the context of the times. It seems to me like Rusch took more cautious consideration of his powerful opponents in his main work, and as a result expressed himself more cautiously but also less comprehensibly than he felt necessary in his very short and precise articles that predominantly appeared in *Kultur und Politik* (*Culture and Politics*). I would thus like to point out that the Zentrum des Organisch- Biologischen Landbaus (Center for Organic-Biological Agriculture) still exists on the Möschberg, and that *Kultur und Politik* is still published there, now in its sixty-fifth year in 2017. (Author's note: The Möschberg is located in the municipality of Grosshöchstetten in the canton of Bern in Switzerland. It is the birthplace of organic agriculture in the German-speaking world. It was founded in 1932 as an agricultural school, and the principal at the time, biologist and high school teacher Dr. Hans Müller, built the foundations of modern organic agriculture together with his wife, Maria Müller-Bigler. Between 1995 and 1996, the building was extensively renovated by Bioforum Schweiz, an organization of organic farmers, and repurposed as an educational institution. The group continues to work toward the strengthening of organic agriculture. The Möschberg contains the text archives from the era of Müller and Rusch. Today the Möschberg functions as an independent seminar-oriented hotel.) Again, all I can do is to repeat myself and appeal once more to researchers and scientists: a large field of research is lying fallow here, and any research or development efforts would be a great help!

QUOTATIONS FROM HUGO SCHANDERL

I would like to conclude by leaving you with some important quotations from Hugo Schanderl (1964) that act as a short, easily understandable summary of what I hoped to convey with this book (the most important passages are highlighted in bold): **"On the emergence of bacteria from plant cells.**

"I dedicated an entire chapter of my 1947 book *Botanische Bakteriologie auf neuer Grundlage* [*A New Basis for Botanical Bacteriology*] to the subject of the bacteria in the soil. The 'new basis' alluded to in the title was the discovery, supported by plentiful experimental proof, that the old idea of 'internal sterility' in the higher plants was mistaken. **I have devised and described many methods and experimental designs that all pointed in the same direction, toward the same result,** namely that fully viable germs can emerge from plant tissue in a form we've long known of and is found all over—bacteria.

"Every modern discovery in molecular biology has supported the idea that mitochondria evolved from bacteria. In many plants, they can still be transformed back into autonomous, culturable bacteria through relatively simple methods. Even larger organelles such as plastids evolutionarily originate from bacteria and can turn back into them under experimental conditions. Most striking of all is the transformation of chlorophyll-generating chloroplasts back into bacteria: the very first bacteria that emerge from them are still visibly green during their first hours of life. The entire process of the remutation of chloroplasts into their original evolutionary forms is 'color coded' by nature. **This disproves the objection that the appearance of the bacteria in this experiment might somehow have been a result of outside infection**.

"The death of a plant gives rise to billions of new lives through the transformation and remutation of organelles into their earlier evolutionary form—bacteria.

"When plant parts or whole plants are buried or composted in the soil, the lives of the particular plants or plant organelles come to an end, but not the life itself. When a plant is buried, the soil is enriched with bacteria not only because a vast number of existing soil bacteria decompose and break down the 'plant corpse,' multiplying tremendously in the process, but also because the soil is enriched with bacteria from higher plants as they break themselves down. Certainly, bacteria present in the soil also find abundant nutrients during composting, which allows them to multiply. But, as can be experimentally demonstrated, no bacteria need to enter from the outside whatsoever for decomposition to take place and a breeding ground of bacteria to arise. **Even while alive, higher plants are already enriching the soil with bacteria**.

"It is well known that the root hairs of all higher plants have short lifespans, ranging from a few hours to a few days. New root hairs form and old root hairs die off every day, and the organelles of those that die can revert back to an earlier evolutionary form, namely that of autonomous bacteria. Because of this, every higher plant provides its surroundings with species-specific bacteria during the course of its life. This is also why the rhizospheres around plants contain more bacteria than more distant areas. **This fact has been known for a long time**, but the explanation given by soil bacteriologists is that the rhizosphere contains more bacteria than other soil because the roots' excretions attract them.

"Every higher plant uses this method to populate the soil around it with species-specific, indigenous bacteria. This is why the soil gets richer in bacteria as more plants are growing in it. The issue of 'soil fatigue' among plants is also related to this. There are crops that react negatively to these species-specific, newly-autonomous former organelles, but demonstrate a clear increase in growth rate under crop rotation. **This promises to be a broad, rewarding area of agricultural research**.

"We do indeed kill the organization of the piece of tissue or the seedling with it, but not the life within it itself. As long as their soil continues to receive enough water, dead plants or tissues sooner or later develop into the same bacteria that we can obtain from their seeds, fruits, or stems via aseptic tissue removal. The same thing happens when farmers plow living plants under or when gardeners compost. The soil is enriched with bacteria that originate in the dead plants. The 'death' of the plant gives rise to billions of new lives, not only because the preexisting soil bacteria can use this new nutrient source to multiply, but also because organelles of the 'dead' plant are resuming their lives in the form of bacteria.

"Plant chlorophyll even gives rise to bacteria in animal rumens! Studying the transformation of chloroplasts is so beneficial because the entire process is 'color coded' by nature. The first bacteria from the chloroplasts are initially an unmistakable green color. As they divide and divide, the fat-soluble coloring fades, and after three or four divisions the bacteria are colorless. I performed related aseptic tissue operations in 1933.

"Depending on the type, it took five to ten days for the chloroplasts to begin to transform into motile bacteria in three distinct, but constantly recurring ways. **I was able to record the entire transformation process from intact chloroplasts into the first motile bacteria on photomicrographs.** I published five of the best photomicrographs as part of my article 'Ein Beitrag zur Frage der Herkunft von Weinbakterien' ['A Contribution on the Question of the Origins of Wine Bacteria'] in the magazine *Weinberg und Keller* [*Vineyard and Cellar*] (volume 16, 1969). In the course of this research, it occurred to me to examine what actually happens to chloroplasts from green plants in the stomachs of ruminants. [. . .] I acquired the rumen of a freshly-slaughtered sheep that was filled with fresh green material. To my amazement, the same transformation of chloroplasts into bacteria took place in the sheep rumen, with the same intermediate stages that I was [. . .] consistently able to observe during my experiments with grape chloroplasts. Studying the rumen's contents [. . .] made it clear to me that processes are taking place in the stomachs of ruminants that we had no idea of before. The bacterial flora in the rumen are constantly being replenished by transformed cell organelles that carry on the biochemical activity of the digested plants. There is thus an intimate relationship between cow and plant, a sort of symbiosis of plant and animal. The cows' feces carry a significant portion of the bacteria regenerated from the plant organelles along with it back to the soil. Unlike artificial fertilizers, this sort of fertilizer is full of life and enriches the soil with bacteria, increasing its fertility. **It has been demonstrated time and time again that fertilizing exclusively with chemical, biologically-dead fertilizer salts eventually makes the soil infertile. The soil becomes impoverished if it doesn't contain any living fertilizer in the form of green manure or ruminant feces.** The farmers of the enormous Chinese civilization have been making use of 'living fertilization' for millennia, putting it into practice with the famous Chinese industriousness. I hope that the agricultural 'development aid' we send to developing countries heeds the fact that agricultural problems in the tropics and subtropics cannot be solved by machines and artificial fertilizers alone. Beneficial effects can only be achieved if you think and work biologically above all else, rather than just economically and technologically.

"Modern bacteriology is still too firmly rooted in the monomorphism of the previous century. It still hasn't begun to think evolutionarily. It views bacteria as its own sort of plant, a type that is evolutionarily very old, but had nothing to do with the evolution of the higher plants.

"However, we must simply consider that plant and animals cells, which have been viewed as the smallest building blocks of life up until now, did not simply appear all of a sudden in nature, but followed their own lengthy evolutionary history before nature began to use them as the components of larger life forms." (Author's note: for more on this subject, see Francé's *Plasmatik* [*Plasmatics*; 1923] and Margulis's *Symbiotic Planet* [1999].) "Over the course of evolutionary history [. . .] primordial cells [. . .] came together into larger cooperative units. The constituent members of these larger cells lost their independence in the union, and were assigned special roles in the new larger cells, eventually becoming functional organs and organelles." (Author's note: see Margulis's "Endosymbiosis" [1999].) "The time will hopefully soon come when these easily-reproducible experiments are incorporated into the curriculum of every practical course on plant physiology. **And the time will soon come that this area of bacteriology will become a fundamental component of the science of cell physiology.** It's true that new discoveries always need time to develop before they are widely accepted. But it seems to me that the amount of time needed in my case shouldn't be necessary anymore in the atomic age."

Section 5

Appendix

Glossary

This book is aimed at farmers, gardeners, and other specialists but also at interested laymen and hobbyists. This glossary defines all of the important technical and biological terms used in this book, including new terms defined or suggested by me, so that it can be accessible to nonexperts as well.

Aerobic: Dependent on oxygen.

Agrobiology: An area of agricultural science focusing on agriculture's biological foundations.

Alternative agriculture: My personal term of choice for the once clearly defined field of organic-biological agriculture, whose foundations I wish to bring back into the spotlight with this book. Along with its predecessors and successors, it is the key to carrying out Hans Peter Rusch's "science of tomorrow." In German, I consciously use the older word *Agrikultur* instead of the more common equivalent *Landwirtschaft* because of its explicit inclusion of "culture."

Amino acids: A group of *organic* acids characterized by a particular arrangement of *nitrogen* in the molecule. Certain amino acids serve as the building blocks for *proteins*.

Amoeba: Relatively large single-celled organisms that constantly change their "body shape." They occur all over the world in the soil as well as in both fresh water and salt water.

Anaerobic: Not dependent on oxygen for metabolism.

Anaerobic decomposition: The breakdown of *organic material* by *microorganisms* in the absence of oxygen.

Anthropogenic: Caused by humans.

Antibiotics: Substances that kill off microorganisms (i.e., living cells). Many organisms naturally produce antibiotics to fight off enemies or disease. When used in medicine, they are synthetically manufactured. In a broader sense, antibiotics are *biocides*.

Antigen: Structure on the surface of a cell (made up of protein or sugar molecules) that a specific antibody (a defensive molecule formed by the immune system) can bind to in order to render the foreign cell harmless.

Archaebacteria: A particular early form of microorganism.

Assimilation: The conversion of absorbed material into material endogenous to the body, or to put it more briefly: the buildup of body matter (the opposite is dissimilation, or the breakdown of nutrients into their components or basic materials).

Atmosphere: The layer of gases surrounding planets and celestial bodies. The Earth's atmosphere consists of 78 percent nitrogen in gas form, 21 percent oxygen, and a total of 1 percent carbon dioxide, noble gases, water vapor, and some other compounds.

Autotrophy, autotrophic (literally, "self-feeding"): The ability of organisms to produce their own body structures and nutrients from *inorganic* material and energy. The best-known examples are plants, which take in *carbon dioxide* from the air and form complex *organic* compounds with the help of their green pigment (*chlorophyll*) and solar energy.

Bacteriophage: A virus that prefers bacteria as host cells and "consumes" them.

Bioavailabilty: The form in which a nutrient is available (e.g., "pure," solid or gaseous, or chemically *bound* in a certain form) to be absorbed and digested by an organism. If a nutrient is present in a non-bioavailable form, it may still be absorbable, but it will not be digestible and it will leave the organism without having been used.

Biocenosis: Collective term for all of the life in a given living area or *habitat*. Not to be confused with an ecosystem, which describes both the biocenosis and the habitat (i.e., both the place and the life within it).

Biocide: A substance that kills undesired living organisms. For example, herbicides (which destroy plants), fungicides (which destroy fungi), or insecticides (which destroy insects).

Biodynamic agriculture: Agriculture in accordance with the anthroposophic principles of Rudolf Steiner, the strictest of the various organic agriculture guidelines. It is practiced and codified by the Demeter organization. Not to be confused with *organic-biological agriculture*.

Biological tillage: See *edaphon*.

Biology: The study of life. What I hope to rediscover along with the reader is "bio-biology"—the life within biology.

Biomass: The total sum of living material, measured in units of weight.

Biomineralization: The formation of structures and frameworks from lime or silicon by *microorganisms*.

Biosphere: Collective term for the areas of the Earth in which life exists.

Blue-green algae: See *cyanobacteria*.

Bokashi: As used in gardening and agriculture, organic plant remains from gardens, fields, and kitchens that have been fermented by effective microorganisms, which increases their quality along similar principles to silage or sauerkraut. Bokashi improves the "feeding" of the life in the soil both biologically and biochemically.

Bound, chemically: A connection between one organic substance and another that causes them to no longer be chemically reactive for as long as they remain bound. Changes to the environment can cause them to become unbound.

C: See *carbon*.

Carbon: A chemical element (symbol: C) that is the hallmark of *organic chemistry* and all *organic material*. All biological molecules are carbon based, often in the form of chain- or ring-shaped compounds.

Carnivore: Meat eater (in contrast to herbivores, or plant eaters).

Catalysis, catalytic: The speeding up of a chemical reaction. Some chemical reactions take place so slowly in the absence of outside influences that the catalysis effectively sets them in motion to begin with. Most metabolic processes within organisms take place catalytically, with *enzymes* serving as the catalysts, for which reason they are also referred to as biocatalysts.

Cellulose: A component of plant cell walls. In chemical terms, it is a *polysaccharide* (complex sugar). Lignifying plants (trees and bushes) build a particularly large amount of cellulose.

Chemoautotrophy, chemoautotrophic: A metabolic system used by certain single-celled organisms in which they do not feed themselves on other organisms but instead live off of (usually very specific) chemical energy (sulfur bacteria are one example).

Chemolithotrophy, chemolithotrophic: A metabolic system used by certain single-celled organisms in which they do not feed themselves on other organisms but instead live off of *inorganic material* (rock-dwelling algae are one example).

Chlorine: A chemical element (Cl in the *periodic table*); a so-called halogen (salt producer). Chlorine is, among other things, a component of table salt (sodium chloride) but also occurs in a wide variety of *organic* compounds (see also *organochloride*).

Chlorophyll: The green pigment in plant cells that makes *photosynthesis* possible.

Chloroplast: The cellular structure or *organelle* in plants where *photosynthesis* takes place.

Chromosome: A type of molecule that contains genetic information (see also *DNA*).

Ciliates: Single-celled organisms whose cell bodies are covered with cilia. They help to move along nutrients. They are present in the soil and in both fresh water and salt water.

Cyanobacteria (literally, "blue bacteria"): A group of bacteria that are proficient in *photosynthesis*. They were once categorized as algae and known as blue-green algae. Cyanobacteria play an important role in soil and water flora.

Cycle of living material: See *living material*.

Cytoplasm: The basic material of a cell inside the cell membrane. Embedded in the cytoplasm are the *nucleus*, the *organelles*, and other cell structures.

Dark-field live blood analysis: A microscopic technique that can be employed in blood studies.

DDT: Dichlorodiphenyltrichloroethane, an insecticide with high potential to cause environmental damage and harm human health. Due to its long half-life, it is accumulating everywhere in the *biosphere*. It has been banned in Germany and many other industrialized countries since the 1970s.

Dead material: My preferred term for the basic components of matter, the chemical elements, as represented in the *periodic table*. I call them dead material as a constant reminder that the "absolute rulership" that we conventionally ascribe to these substances is not sufficient to solve our ecological problems.

Deep ecology: A natural philosophy and ecological movement founded by the Norwegian philosopher Arne Nœss (1912–2009) that Lynn Margulis and James Lovelock incorporated into their *Gaia hypothesis*.

Derivative: A chemical substance that is derived from a different one or is formed through minor changes to the individual atoms in the molecule. A derivative can have dramatically different chemical and physical characteristics from the original substance.

Diatoms: Single-celled algae that are present in fresh water, salt water, and the soil.

Dissociation: Breakdown (e.g., of a molecule into its component parts).

DNA: Deoxyribonucleic acid, the molecular building block that forms the genetic material in living beings (e.g., *chromosomes*, *genes*). Recent discoveries show that genetic material is present not only in cell nuclei in the form of chromosomes but also in other parts of the cell, where it takes on special functions (*epigenetics*).

Ecological agriculture: A collective term for agriculture and livestock production that is done without pesticides, without *mineral fertilizers*, and without genetic modification. The term also encompasses *biodynamic agriculture*, which is a more specific subtype.

Ecology (from the Greek *oikos*, meaning "house" or "household"): The study of natural balance. Unfortunately, this is a somewhat overused term nowadays. I'd like to refer here to Fritjof Capra (1994), who "gave a rigorous diagnosis [. . .] of how modern man has forgotten that reality is not as real as it seems, that everything is a constantly changing dynamic process. [. . .] In the natural sciences [. . .], the theory of living systems that has developed over the last decades provides the ideal framework for formulating a new type of ecological thought. I therefore use the terms "ecological" and "systemic" synonymously.

Edaphic: See *edaphon*.

Edaphon: A collective term for the small life-forms living in and on the soil (e.g., fungi and other microorganisms, plants, and animals). It was defined in 1911 by Raoul H. Francé as "a biocenotic living community along the same lines as plankton; in a certain sense the 'plankton of the soil.'" Unlike water, which maintains its structure even in the absence of plankton, the structure of the soil, the humusphere is entirely dependent on edaphon. Margaret Sekera (1984) described this as "the biological tillage of the soil," and her husband, the soil scientist Franz Sekera, called it "the biological tillage of the crumb structure" (Sekera 1951). The *biosphere's* collective metabolism runs through the *humusphere* and is constantly being carried out by the edaphon.

Effective microorganisms: In the early 1980s, Japanese scientists developed a mixture of genetically unmodified microorganisms that represent a promising hope for our depleted, over-fertilized, and poisoned *biosphere* and are now known as effective microorganisms. The scientists mixed together a variety of different microorganisms, consisting of several types of photosynthetic bacteria, lactic acid bacteria, yeasts, and fermentation-capable fungi. Many of these microorganisms have already been used for centuries to varying extents in agriculture and gardening, environmental remediation, medicine, and the food industry and they are very valuable to humans, plants, soil, and water.

There are so many ways to make use of effective microorganisms that they can be employed in practically any area of life. The variety of microorganisms and their adaptability have made it possible to develop a number of compounds with special functions. Since this technology originates in agricultural science, it is not surprising that it is most heavily emphasized within the field, but it is also applied in many different everyday situations. Effective microorganisms are becoming more and more important in some of the areas where they are employed, such as soil and environmental remediation, wastewater treatment, and waste management. There is also an ever-increasing number of foodstuffs, supplements, and care products for humans and animals that are made from or incorporate effective microorganisms.

Systematically introducing starter cultures made from effective microorganisms into the soil, garden plants, orchards, and farm fields causes the plants to grow demonstrably more healthily and to last longer after being harvested. Combined with sufficiently large quantities of organic material, which accrues in large amounts in gardening and domestic chores as well as in farming, the effective microorganisms speed up humus formation in the soil and stabilize it in a sustainable form. Effective microorganisms are the ideal "food base" for plant *endocytosis*.

Endocytosis: A process through which a cell absorbs molecules or particles into its body by enveloping them with its cell wall and then transporting the resulting vesicle into the cell and pinching it off (see also *exocytosis*).

Endosymbiotic theory: A theory that postulates that *organelles* (particular components of a cell) were once single-celled organisms that were eventually absorbed and integrated into cells and carry out specific tasks there (see also *chloroplast, mitochondria*). The basis of the theory is the organelles' form and the fact that they contain their own genetic material (independent of the cell nucleus).

Enzyme (formerly known as a ferment): A substance that speeds up a chemical reaction or sets it in motion to begin with (*catalyzes* it). Most enzymes are large, complex *protein* molecules. All metabolic processes in living organisms are dependent on many different enzymes.

Epigenetics: A field of research concerned with the genetic material (i.e., the *DNA*) that is not located within the nucleus but in other parts of the cell, such as the *organelles*.

Exocytosis: The *endocytosis* process in reverse. Material that the cell wants to release to the outside or get rid of is enveloped in cell wall material. The resulting vesicle is transported to the cell wall, its own walls bind to it, and finally the cell wall opens and releases it outside. This ejects the particles or molecules from the cell while simultaneously closing the cell wall again behind them.

Facultative: Not obligatory; when needed; by choice.

Ferment: See *enzyme*.

Fermentation: The breakdown of organic material by microorganisms in the absence of oxygen. The distinction between fermentation and *anaerobic decomposition* is that fermentation produces energy.

Flagellate (from the Latin *flagellum*, "whip"): A single-celled organism with whiplike features that are generally used for movement. Flagellates can be plants (phytoflagellates) or animals (zooflagellates).

Flood culture: See *hydroculture*.

Gaia hypothesis: A theory developed by biologist Lynn Margulis and biophysicist James Lovelock in the 1960s that postulates that the Earth and its entire collective biosphere should be viewed as a single dynamic, self-regulating, ever-developing organism.

Gene: A segment of a *chromosome* that codes for a single specific piece of genetic information and can be read as needed.

Gene transfer: The transmission of genetic information (i.e., *genes*) from one organism to another. Vertical gene transfer is the transfer from a parental generation to a filial generation, as occurs in reproduction. Horizontal gene transfer refers to the transfer of genes from one organism to another, which can even occur over species boundaries.

Genesis: Appearance, formation, or birth.

Glacial milk: Meltwater from glaciers containing very fine particles rubbed off from the rock (true *minerals*), giving it a milky or cloudy coloring.

Habitat: Living area (see also *biocenosis*).

Hay infusion: A method of "producing" microorganisms. Dried grass and water are placed in a closed container and left standing for a matter of days. This leads to the growth of large quantities of microorganisms.

Heterotrophic: Subsisting on *organic material*, and thus on other organisms.

Higher plants: Multicellular plants with a complex structure of roots, stems, and blossoms, as opposed to the more simply structured lower plants, which include algae (which can also be single-celled), mosses, and ferns as well as the so-called clubmosses. The terms *"endocytosis," "endosymbiotic theory,"* and "plant nutrition" refer to higher plants in the context of this book.

Humus: The *organic* portions of the soil (i.e., the layers that contain biological material). This is always the soil's uppermost layer, which is often just a few centimeters thick. Under normal, natural circumstances (depending on climate and environmental conditions), it can take thousands of years for 10 centimeters of humus to form.

Humusphere: The sphere that all of the *biosphere's* metabolic processes pass through. It is located between the *atmosphere* and the *lithosphere*. It also includes the *hydrosphere*. The humusphere is where metabolic and plant nutrition processes (i.e., all of the "agricultural processes") take place. The search continues for the "dead" components of what is known as the humus, which can only be described in chemical terms. It is best described, however, as "the result of the functions of the edaphon" (Hennig 1994) or perhaps better still as a result of the functions of the humusphere, which leaves the question quite open of whether searching for pure materials can even lead to the most useful explanatory model. I see the humusphere in terms of a self-regulating "functional community" that is driven by living organisms, an approach that has great potential to increase our understanding of the processes taking place in plants and in the soil. The core principles of the *Gaia theory* are helpful here. The *edaphon* is collectively the builder and caretaker of the *cycle of living material*, while at the same time being a direct participant in it. The edaphon fuels all conversions of material—and not by just producing *salt* ions for plant nutrition but as an active participant in the cycle of eating and being eaten. Plants, with their edaphon-eating roots (*endocytosis*), are most certainly a part of this cycle as well.

Hydroculture: One of many historical agricultural systems in which water is emphasized (e.g., cultivation in flood areas or the intentional introduction of water to the growing plants).

Hydrosphere: Collective term for all the Earth's water resources.

Indicator: A material, plant, or substance that serves as evidence for something. For example, heavy stinging nettle growth is an indicator of high nitrogen content in the soil since stinging nettles particularly require it. The presence of lactic acid is an indicator of *anaerobic* metabolic processes or *fermentation*.

Indigenous: Original or native; originating from the place where it was discovered.

Infusoria cultures: The "creation" of microorganisms, such as with a *hay infusion*. They are used, among other things, by fish owners to create food for their fish.

Inorganic chemistry: Covers (with a few exceptions) all matter that does not contain *carbon* (see also *organic*).

Ion: An electrically charged ion or molecule.

Ion exchange: A technical process through which *ions* are exchanged for other ions with the same electrical charge. Some water filters work via the ion exchange principle.

K: See *potassium*.

Lactic acid bacteria: A group of microorganisms that do not require oxygen for their metabolic processes and that produce lactic acid in the course of their chemical reactions, a process known as *fermentation*.

Legume: A family of plants with blossoms that are shaped like butterflies and that produce especially protein-rich seeds. Many of our food plants (e.g., beans, peas, soybeans) belong to this family due to the latter characteristic—the protein content provides a high level of nutritional value. Another property of legumes is that they accumulate *nitrogen*-loving microorganisms known as *rhizobia* in their *root nodules*, which leads to them often being intentionally planted in order to enrich the soil's nitrogen content (i.e., green manure).

Lithobiont (literally, "stone dweller"): Organisms that are capable of living on bare rock.

Lithosphere: The outermost rocky layer of the planet Earth.

Living material (according to Rusch): All plant nutrient material that is alive, such as *microorganisms* or *organelles*, as opposed to *salts*, *minerals*, or molecules, which are indeed building blocks of life but are not themselves alive in the biological sense.

Metabolite: A molecule that forms as an intermediate stage in a chain of biochemical reactions (e.g., during the exchange of cellular material).

Micrococci: A particular group of bacteria.

Microorganisms: Very small life-forms that are invisible to the naked eye. Types of microorganisms include bacteria, algae, and fungi, and most are single-celled (see also *plankton*).

Microsomes: Fragments of the membranes of certain cell components.

Mineral: The building blocks of the Earth's crust and of rocks, primarily composed of *inorganic* compounds. The word "mineral" and all of the terms that incorporate it originate in geology. This book makes sure to use the word "mineral" only in this proper, technically correct sense.

However, I should again mention that it is very widely used incorrectly as well. The modern inaccurate use of the terms "mineral" and "mineralization" within agriculture dates back to Justus von Liebig, who discovered "mineral components" in his ash analyses 170 years ago. But Liebig neglected to work out the distinction between "mineral components" and "minerals" and postulated that pure minerals were plants' source of nutrients.

This lack of precision has been perpetuated through the present day. But what Liebig and his chemists found back then were just the chemical elements (i.e., the contents of the *periodic table*). From a chemical-analytical point of view, these elements are the building blocks that all matter (both living and nonliving) is composed of. In and of themselves, they are neither nutrients nor minerals but rather components of both.

According to the more nuanced definition, only rock, sand, clay, stone meal, volcanic ash, and the mineral components of glacial milk can be described as minerals. Everything else needs a different term.

Mineral fertilizers (also known as NPK fertilizers, with the "N" standing for *nitrogen***, the "P" for** *phosphorus***, and the "K" for** *potassium***):** Fertilizers composed of *inorganic* compounds, also known as *salts*. The *nitrogen* used in them is produced through large-scale industrial manufacturing from oil and is used to manufacture explosives in addition to the fertilizer.

Mineralization: When substances mineralize, they turn into *minerals*—fossilized plants and animals or deposits in the Earth's crust. In contrast to this original geological definition, the term is used in agrochemistry to describe the breakdown process of primarily *organic material* (i.e., its *dissociation* in watery solutions). But this conceptualization is unduly simplified. It disregards all of the other processes taking place in the *humusphere*, or in a worst-case scenario even labels them as unscientific. Modern agricultural doctrine and the practice of plant nutrition limit themselves to this excessively simplified and thus effectively distorted conceptualization.

Mineral theory: The theory that plants feed exclusively on *minerals* (by which *salts* are actually meant) and not on living cellular material and that they must therefore be supplied with *mineral fertilizers*.

Mitochondria: Small, spheroid structures that are a component of nearly all living cells and contain their own genetic material (see also *endosymbiotic theory*, *organelle*). Mitochondria are often described as the "factories of the cell," as the primary thing that happens within them is energy metabolism.

Monosaccharide: A simple sugar (i.e., a sugar compound that is made up of only a single sugar molecule, such as glucose or fructose). Disaccharides and polysaccharides, such as lactose, starch, or *cellulose*, consist of two or more sugar molecules.

Mulch: A covering for the soil for the purpose of protecting it and preventing erosion, drying, or frost damage to young plants or suppressing the growth of unwanted plants. Mulching using the methods described in this book; however, mulch primarily serves the purpose of providing food for the *edaphon*.

Mycorrhiza: A form of *symbiosis* between fungus and plant in the (fine) root area of the plant. The fungus supplies the plant with nutrients and water from the soil, and the plant in turn supplies the fungus with nutrients from its *photosynthetic* processes, which fungi cannot produce themselves due to a lack of *chlorophyll*. The best-known form of mycorrhiza, which can also be very specific in nature, is the growth of Agaricomycetidae mushrooms under certain trees.

N: See *nitrogen*.

Nematode: A very large phylum in the animal kingdom consisting of more than twenty thousand species. They are found in both the soil and water. Some nematode species are parasitic or pathogenic, such as intestinal roundworms.

Nitrogen: A chemical element (symbol: N) and component of NPK *mineral fertilizers*. Also an important component of *proteins*.

Nitrogen availability: An indication of what chemical form *nitrogen* is taking in the soil (as nitrogen gas, in *ion* form, or bound within molecules). This determines whether the nitrogen can be absorbed and utilized by the plant.

Nitrogen harvest index (NHI): The amount of a given nutrient (usually nitrogen) in a harvested product as a proportion of the total amount of the nutrient in all above-ground biomass.

NPK: See *mineral fertilizers*.

Nucleus: A separated area in the interior of a cell that contains its genetic material (*DNA*).

Oligosaprobes: A particular group of organisms that live off dead *organic material* (see also *saprophytes*).

Opportunistic: Describes an organism that is highly adaptive to changing environmental conditions.

Orbio: Shortening of organic-biological. It was under this name that the work of Hans Müller and Hans Peter Rusch was introduced and spread in Scandinavia by Hans Cibulka from Sweden and Ivar Torp from Norway. The same methods are promoted by the Bioland-Verband in Germany.

Organelle: Small bodies within a cell. They form closed-off regions in which certain metabolic processes take place. *Mitochondria*, the so-called factories of the cell where energy is supplied, are an example of one type of organelle. *Photosynthesis* takes place in the *chloroplasts* of plant cells, another type of organelle.

Organic-biological agriculture: An ecologically based agricultural system that arose from the reform movement of the nineteenth century. It is based on the discoveries of the microbiologist Hans Peter Rusch. Not to be confused with *biodynamic agriculture*.

Organic chemistry: The subfield of chemistry that deals with *carbon*-based substances. *Organic material* is the hallmark of living organisms (see also *inorganic chemistry*).

Organic material: Any substance that is built around and primarily composed of the chemical element *carbon*. All other substances are described as *inorganic*.

Organochloride: Substance consisting of a *carbon* skeleton (meaning it is organic) along with one or more *chlorine* atoms. Many *biocides* are or contain organochlorides. Most of these compounds are harmful to your health and to the environment (e.g., *DDT*).

Organophosphorus: *Carbon*-containing (i.e., *organic*) chemical compounds that contain one or more phosphorus atoms.

P: See *phosphorus*.

Pathogenic: Disease-causing.

PCB: Polychlorinated biphenyl. A group of chemical compounds (*organochlorides*) that are used, among other things, as plasticizers and in electrical engineering. Due to their extreme toxicity and potential to cause cancer, they are among the "dirty dozen" highly toxic chemicals that have been banned worldwide by the Stockholm Convention on Persistent Organic Pollutants.

Periodic table: A representation of the chemical elements sorted by their atomic weights and electron configurations.

Permaculture: A combination of the words "permanent" and "agriculture." Its goal is to combine various different ecological methods on scales ranging from a single house or a small settlement all the way to large swathes of land. Permaculture translates the principles of a closed ecological cycle in nature to the planning of human settlements. The movement was sparked by two Australians, Bill Mollison and David Holmgren, in 1978. Possibly the best definition of permaculture comes from Bill Mollison himself: "Permaculture is a dance with nature—in which nature leads."

Pesticide: A chemical substance used to ward off undesired plants or animals by inhibiting them or by killing them (see also *biocide*).

pH value: Unit of measure of a substance's acidity (or the soil's). A pH value of 7 is defined as neutral, while values under and over 7 correspond to more acidic and more alkaline environments respectively.

Phosphorus: A chemical element (symbol: P) that is important to living organisms as, among other functions, a component of genetic material (*DNA*) and in the metabolism of fat and energy. Plants need phosphorus to conduct *photosynthesis*. It is a component of *mineral fertilizers* (*NPK*).

Photoautotrophy, photoautotrophic: The characteristic metabolism form used by plants. They do not feed on other living organisms, but rather absorb energy from sunlight (*photosynthesis*). According to the *cycle of living material* theory, however, it is not in fact the sole form of plant nutrition.

Photosynthesis: The process through which plant cells, with the help of sunlight, build up living material, including their own bodies, out of water and atmospheric CO_2 (*carbon dioxide*). Photosynthesis takes place in *chloroplasts*.

Phylum: A term from the taxonomic system.

Plankton: Collective term for the smallest creatures living in the water (both fresh water and salt water). Plankton does not actively move; it floats along with the current. Plankton consists of both animal (e.g., tiny crabs) and plant (e.g., algae) organisms and acts as the main source of nutrients for many aquatic animals (e.g., fish, baleen whales). Soil water also contains plankton (*edaphon*).

Polysaccharide: See *monosaccharide*.

Polysaprobes: A particular group of organisms that live off of dead *organic material* (see also *saprophytes*).

Pore volume: The sum total of the pores in the soil, expressed as a proportion of the total volume. Pore volume plays an important role in soil quality because it is responsible for aeration and gas exchange.

Potassium: A chemical element (one of the alkaline earth metals) and a component in *NPK* mineral fertilizers.

Primary producer: An organism that is not dependent on other organisms as food sources, but can instead build up its body out of *inorganic* material, making it available to other organisms as food (like plants through *photosynthesis*).

Proteins: Large, complex molecules that always contain one or more atoms of *nitrogen* in a particular configuration. Proteins are the building blocks of many organic molecules, including genetic material (*DNA*), *enzymes*, and hormones. Their own building blocks are *amino acids*.

Protoplasm: Obsolete term for *cytoplasm*.

Reductionist (from the Latin *reducere***, "to lead back"):** A way to describe a doctrine or system of thought that reduces any phenomenon to its smallest possible components or building blocks and only considers it from that perspective.

Regeneration: Restoration (e.g., of an original state).

Remutation, remutate: According to the *cycle of living material* theory, the ability of cell *organelles* to revert to their original form as viable single-celled microorganisms after the death and decomposition of the cell.

Rhizobia: A particular group of soil-dwelling bacteria that are capable of binding to *nitrogen* in gas form from the air (within the soil). Plants from the *legume* family—including peas, beans, clover, and lupines—can accumulate rhizobia in their roots. This *symbiosis* is beneficial to the plant, for which nitrogen is an important nutrient.

Rhizopods (literally, "root-footers"): A particular group of single-celled organisms.

Rhizosphere (literally, "root space"): The area of the soil surrounding plant roots.

Root hairs: The smallest outcroppings of the outermost plant root cells. Root hairs are very numerous and increase the plant root's surface area enormously, which makes absorbing nutrients easier. According to the *cycle of living material* theory, root hairs also absorb large molecules and even entire cells (e.g., **microorganisms**) as nutrients via *endocytosis*.

Root nodule: A formation of the roots of certain plants that contains *rhizobia*.

Rotifers: Tiny organisms with a ring of cilia on the head ("wheel organs") that helps with both moving and rotating food sources. Rotifers are found among *plankton* as well as in the soil and other habitats. They can be very highly specialized.

Rotting: A process in *humus* formation where dead plant material is broken down and returned to the soil. The process is called decomposition when describing animals.

Salt: A chemical compound made up of positively and negatively charged *ions*. Due to the electric attractive force, they form regular crystal structures. There are both *organic* and *inorganic* salts.

Saprophytes: Organisms that live off other dead organisms. Saprophytes are important in the decomposition of dead animals and plants in the soil.

Semipermeable: Only permeable to a portion of the substances or molecules.

Semiochemical: A chemical substance in an organism that helps with transferring signals and cellular communication.

Solution, soluble: Any material that dissolves in a watery liquid is simultaneously forming a "mixture" with the water molecules. The result is still a liquid, and while its color or clarity may change, it will remain homogenous, or uniform. For example, sugar or salt can be dissolved in water, but sand cannot. Oil and fat are also not water soluble; they settle in the water instead. They are, however, soluble with each other (so you can mix together olive oil and sunflower oil homogenously, for example).

Spontaneous generation: See *genesis*.

Stem cell: A cell that has not (yet) been differentiated and is capable of developing into one of multiple particular types of cells (e.g., into a liver cell or a leaf cell). Stem cells occur naturally in certain developmental stages of both plants and animals and also to a certain extent in already-differentiated tissue.

Symbiont: A partner in a *symbiotic* relationship.

Symbiosis: A mutually beneficial relationship between two (or more) organisms, usually of different species, in which both are dependent on each other. Lichens are an example of symbiosis in which an alga and a fungus "lived together" so closely that a new life-form emerged. Other examples of symbiosis include human digestive and skin flora or the *edaphon* living in plant root areas. Our biosphere is filled with symbiotic life-forms.

Symbiosis control: A therapy in alternative medicine with the goal of restoring the proper community of natural digestive bacteria.

Symbiotic: See *symbiosis*.

Technological agriculture: The currently predominant form of agriculture, based on the *mineral theory* and making use of chemical and mechanical methods.

Thousand kernel weight (TKW): Also known as thousand grain weight (TGW). Parameter for calculating seed quantities.

Uniform: Consistent; always taking the same form or shape.

Waiting period: The prescribed period, according to the instructions, that should elapse between when a *pesticide* is applied and the harvest. The presumption is that the substances will be so thoroughly broken down or will have so thoroughly seeped away by this point that the legally established limits on them in the resulting fruits, vegetables, or grains will not be exceeded. It does not mean that the food will be free of pesticides by that point.

Yeast: A group of single-celled fungi that reproduce by sprouting (forming daughter cells).

Zoogloea: A group of single-celled organisms that take the form of slimy layers or coatings.

Bibliography and Further Reading

Åkersted, N. (1993): Boken om marktächning och om odling i sand. Natur och Trädgård, S61621 Äby 1995 (Only availabe in Swedish)

Armstrong, Karen (2007): Eine kurze Geschichte des Mythos. dtv Verlag, Munich
Balzer, Fritz M. (1999): Ganzheitliche standortgemäße dynamische Bodenbearbeitung. Verlag Ehrenfried Pfeiffer Ausbildungsund Forschungsstätte, Wetter

Balzer-Graf, Ute and Balzer, Fritz (1991): Steigbild und Kupferchloridkristallisation – Spiegel der Vitalaktivität von Lebensmitteln –. In Meier-Ploeger, A. M., Vogtmann H., (Editor): Lebensmittelqualität – ganzheitliche Methoden und Konzepte. Verlag

F. Müller, Karlsruhe, 2nd edition, pages 163–210

Batmanghelidj, F. (1997): Wasser – die gesunde Lösung. VAK Verlag, Breisgau
Bayerische Akademie der Wissenschaften (Editor): Bedeutung der Mikroorganismen für die Umwelt. Rundgespräche der Kommission für Ökologie, volume 23. Verlag Dr. Friedrich Pfeil, Munich

Béchamp, Pierre Jacques Antoine: (1883): Les microzymas dans leur rapports avec l'hétérogénie, l'histogénie, la physiologie et la pathologie. Examen de la panspermie atmosphérique continue ou discontinue, morbifère ou non morbifère. Baillière, Paris. Neuauflage durch das Centre International de Recherches Antoine Béchamp (1990) *Beste, Andrea et al.* (2001): Bodenschutz in der Landwirtschaft. Einfache Bodenbeurteilung für Praxis, Beratung und Forschung. Stiftung Ökologie & Landbau, Bad Dürkheim

Blech, Jörg (2000): Leben auf dem Menschen. Die Geschichte unserer Besiedler. Rohwolt Verlag, Reinbeck

Blotzheim, von Urs N. (1993): Handbuch der Vögel Mitteleuropas. volume 13/1, pages 791 ff. Aula Verlag, Wiebelsheim

Botsch, Walter (1975): Entwicklung zum Lebendigen. Die chemische Evolution. Franckh´sche Verlagshandlung W. Keller, Stuttgart

Brauner, Heinrich (2010): Die Grundlagen des organisch-biologischen Landbaus – wie sie von den Pionieren Dr. Hans und Maria Müller und Dr. Hans Peter Rusch erarbeitet worden sind. Herausgegeben von der Fördergemeinschaft für gesundes Bauerntum, A-4020 Linz

Brettcher, Mark S.: How animal cells move. In: Scientific American, 255, 72–90 (1987)

Buch, Walter (1986): Der Regenwurm im Garten. Ulmer Verlag, Stuttgart

Buhlert, Hans (1902): Untersuchungen über die Arteinheit der Knöllchenbakterien in den Leguminosen und über die landwirtschaftliche Bedeutung dieser Frage. Habilitationsschrift, Friedrichs-Universität Halle-Wittenberg

Capra, Fritjof (1982, 1994): Wendezeit. Bausteine für ein neues Weltbild. Scherz Verlag, Bern

Carson, Rachel (1962), (1990): Der stumme Frühling. Beck Verlag, Munich

Caspari, Fritz (1948).: Fruchtbarer Garten, Heering-Verlag, Seebruck

Chaboussou, Francis (1985): Pflanzengesundheit und ihre Beeinträchtigung. Verlag C.F. Müller, Karlsruhe

Cibulka, Hans (1999): Kontakt, Undervisningsorgan för Organisk Biologisk Odlingskultur in Skandinavien. 30. Ärgäng

Claro, Elisa (1999). Specific source no longer available

Coats, Callum (1999): Naturenergien verstehen und nutzen. Viktor Schaubergers geniale Entdeckungen. Omega Verlag, Düsseldorf

Dahl, Jürgen (1989), (1996): Die Verwegenheit der Ahnungslosen.

Über Genetik, Chemie und andere Schwarze Löcher des Fortschritts. Klett Cotta Verlag, Stuttgart

Darwin, Charles (1881), (1983): Die Bildung der Ackererde durch die Thätigkeit der Würmer. März Verlag, Berlin und Schlechtenwegen

Dittmann, Jürgen and *Köster, Heinrich.* (2004): Die Becherlupenkartei mit Begleitheft: Tiere in Kompost, Boden und morschen Bäumen. Verlag an der Ruhr, Mülheim *Dixon, Bernard* (1998): Der Pilz, der John F. Kennedy zum Präsidenten machte – und andere Geschichten aus der Welt der Mikroorganismen. Spektrum Akademischer Verlag, Heidelberg

Dunger, Wolfram (1983): Tiere im Boden. Westarp Wissenschaften, Hohenwarsleben *Dürr, Hans-Peter* (2011): Das Lebende lebendiger werden lassen. Wie uns neues Denken aus der Krise führt. Oekom Verlag, Munich

Dürr, Hans-Peter (2009): Warum es ums Ganze geht. Neues Denken für eine Welt im Umbruch. Oekom Verlag, Munich

Dunbar, William Phillips (1907): Zur Frage der Stellung der Bakterien, Hefen und Schimmelpilze im System. 2. Edition, erweitert mit einer Einführung von Georg Meinecke 1981. Semmelweis-Institut, Hoya. (Contains very important information on Günther Enderlein and Hugo Schanderl under the keyword"Pleomorphismus" ["Pleomorphism"])

Emoto, Masaru (2010): Die Botschaft des Wassers. Sensationelle Bilder von gefrorenen Wasserkristallen. Koha Verlag, Burgrain

Endlich, Bruno et al. (1985): Der Organismus der Erde. Grundlagen einer neuen Ökologie. Verlag Freies Geistesleben, Stuttgart

Engqvist, Magda (1977): Die Steigbildmethode. Ein Indikator für Lebensprozesse in der Pflanze. Vittorio Klostermann, Frankfurt am Main

Erven, Heinz (1981), (2014): Mein Paradies. 32jährige Erfahrungen eines Praktikers im naturgemäßen Obstund Gartenbau. OLV Verlag, Kevelaer

Feist, Ludwig (1954): Garten, Heim und Gewinn. Verlag der Raiffeisendruckerei, Neuwied

Feist, Ludwig (1948): Der Familiengarten Seine wirtschaftliche und soziale Bedeutung. Verlag Br. Sachse, Hamburg.

Filser, J. (1997): Der Boden lebt. In: GSF Forschungszentrum für Umwelt und Gesundheit (Editor): Böden – verletzliches Fundament. In: mensch + umwelt, 11th edition, pages 43 – 49, Neuherberg

Francé, Raoul Heinrich (1911), (2012): Das Leben im Boden/Das Edaphon. Untersuchungen zur Ökologie der bodenbewohnenden Mikroorganismen. OLV Verlag, Kevelaer

Francé, Raoul Heinrich (1923): Plasmatik – Die Wissenschaft der Zukunft. Walter Seifert Verlag, Stuttgart/Heilbronn

Francé-Harrar, Annie (1964),(2014): Leben aus dem Stein. Der mikrobiologische Ursprung des Humus und seine ständige Neuschöpfung durch die Lithobionten. (A new edition annotated by Dr. Ines Fritz, Universität für Bodenkultur (BOKU), Vienna, to reflect the current state of the field was release in 2014 as an eBook. Universität für Bodenkultur, Vienna.)

Francé-Harrar, Annie (1959), (2011): Handbuch des Bodenlebens. BTQ e.V., Kirchberg und Blue Anathan Verlag, Haigerloch

Francé-Harrar, Annie (1950), (2007): Die letzte Chance. Für eine Zukunft ohne Not. BTQ e.V., Kirchberg und Blue Anathan Verlag, Haigerloch

Francé-Harrar, Annie (1957): Humus. Bodenleben und Fruchtbarkeit. Bayerischer Landwirtschaftsverlag, Munich

Frank, Albert Bernhard (1890): Über die Pilzsymbiose der Leguminosen. Sonderdruck aus den Landwirthschaftlichen Jahrbüchern. Paul Parey Verlag, Berlin

Füller, Horst (1954), (2002): Die Regenwürmer. Verlags KG Wolf, Magdeburg
Fukuoka, Masanabu (1975),(2013): Der Große Weg hat kein Tor. pala verlag, Darmstadt

Fukuoka, Masanabu (1987), (1999): Die Suche nach dem verlorenen Paradies. Natürliche Landwirtschaft als Ausweg aus der Krise. pala verlag, Schaafheim

Fukuoka, Masanabu (1998): Rückkehr zur Natur. Die Philosophie des natürlichen Anbaus. pala verlag, Schaafheim

Fukuoka, Masanabu (1988), (1998): In Harmonie mit der Natur. Die Praxis des natürlichen Anbaus. pala verlag, Schaafheim

Furst, P. T. (1978): Spirulina, a nutritious alga, once a staple of Aztec diets, could feed many of the world's hungry people. In: Human Nature, March 1978, pages 60–65

Glenk, Wilhelm and Neu, Sven (1990): Enzyme. Die Bausteine des Lebens – Wie sie wirken, helfen und heilen. Wilhelm Heyne Verlag, Munich

Görlich, J.J. (no year, ca. 1940s): Fruchtbare Erde – Ein Wegweiser der beschleunigten Humuserzeugung und Bodenbefruchtung mit Regenwurmkulturen. Eigenverlag. *Graff, Otto* (1958): Die Regenwürmer Deutschlands, M & H. Schaper Verlag, Hanover

Graff, Otto (1983): Unsere Regenwürmer. Verlag M. & H. Schaper, Hanover
Haller, Albert von (1983): Macht und Geheimnis der Nahrung. Bioverlag Gesund leben, Hopfenau

Haller, Albert von and Haller, Wolfgang von (1976, 1978): Die Wurzeln der gesunden Welt I und II. Verlag Boden und Gesundheit, Langenburg

Haller, Albert von (1977): Lebenswichtig aber unerkannt. Phytonzide schützen das Leben. Verlag Boden und Gesundheit, Langenburg

Hamm, Wilhelm (1872), (2002): Das Ganze der Landwirtschaft in Bildern. Arnoldische Buchhandlung Leipzig. Unchanged"libri rari" reprint edition, Verlag Th. Schäfer im Vincentz Verlag, Hanover

Hartmann, A. (1997): Verborgene Welt im Kleinen. In: GSF – Forschungszentrum für Umwelt und Gesundheit (Editor): Böden – verletzliches Fundament. In: mensch + umwelt, 11th edition: pages 50 – 56, Neuherberg.

Haye, Erich (1993): Das ist der Mann, der vieles kann: Siegfried Lange in Felbecke. (Die Sägemehltheorie.) In: ANDERS LEBEN, o. Jg., volume 42, pages 2-5

Haye, Erich (1985): Jedes Beet eine Gesellschaft für sich. Serie über einen erfolgreichen Versuch im Hochsauerland. (Die Sägemehltheorie.) In: ANDERS LEBEN, o. Jg., volume 4, pages 96-99

Haye, Erich (1985): Gemüse zwischen dem Getreide. Fortsetzung des Artikels "Jedes Beet eine Gesellschaft für sich". (Die Sägemehltheorie.) In: ANDERS LEBEN, o. Jg., volume 5, pages 131-134

Heinrich-Böll-Stiftung et al. (Editor), (2015): Bodenatlas. Daten und Fakten über Acker, Land und Erde. Heinrich-Böll-Stiftung, Berlin

Hellriegel, Hermann and Wilfarth, Hermann (1888): Untersuchungen über die Stickstoffnahrung der Gramineen und Leguminosen. In: Beilagenheft der Zeitschrift des Vereins der Rübenzuckerindustrie des Deutschen Reichs. Volume 38, Berlin

Hemleben, Johannes (1978): Das haben wir nicht gewollt. Sinn und Tragik der Naturwissenschaft. Verlag Urachhaus & Geistesleben, Stuttgart

Hennig, Erhard (1999),(2017): Kompost in einhundertjähriger Entwicklung. Von den Anfängen der Stallmistbehandlung bis zur perfekten Kompostierung. Eine Dokumentation. OLV Verlag, Kevelaer

Hennig, Erhard (1992),(2017): Humus als Grundlage der Waldernährung – Humusbeschaffung/Humuserhaltung; Rekultivierung des Waldbodens; Ursachen und Auswirkungen des Waldsterbens. OLV Verlag, Kevelaer

Hennig, Erhard (1995), (2017): Geheimnisse der fruchtbaren Böden. Die Humuswirtschaft als Bewahrerin unserer natürlichen Lebensgrundlagen. OLV Verlag, Kevelaer

Hennig, Erhard (1988), (2017): Humustrilogie – Der Weg der deutschen Landwirtschaft; Die Neuordnung gesundes Lebens beginnt beim Humus; Verbindungen der Gesundheit des modernen Menschen mit der Gesundheit des Bodens. OLV Verlag, Kevelaer

Hensel, Julius und Schacher, John (1898), (1939), (2010): Brot aus Steinen (durch mineralische Düngung). 1894 German back-translation. Verlag A.J. Tafel, Philadelphia, PA, USA

Heyer, von Gustav (1973): Die Revolution beginnt im Garten. Eigenverlag, Hamburg *Higa, Teruo* (2000), (2002): Die wiedergewonnene Zukunft. OLV Verlag, Xanten *Higa, Teruo* (1994), (2000): Eine Revolution zur Rettung der Erde. Mit Effektiven Mikroorganismen (EM) die Probleme unserer Welt lösen. OLV Verlag, Xanten *Hitschfeld, Oswald* (1984),(2009): Der Kleinsthof und andere gärtnerisch-landwirtschaftliche Nebenerwerbsstellen. Ein sicherer Weg aus der Krise. OLV Verlag, Xanten, 7th edition

Hoffmann, Heide (1999): Urbane Landwirtschaft am Beispiel der Organoponics in Havanna/Kuba. Deutscher Tropentag 1999 in Berlin, Humboldt-Universität zu Berlin, Landwirtschaftlich-Gärtnerische Fakultät

Hoffmann, Manfred (2007): Lebensmittelqualität und Gesundheit. Bio-Testmethoden und Produkte auf dem Prüfstand. Barens & Fuss, Schwerin

Hoffmann, Manfred (1997): Vom Lebendigen im Lebensmittel. Deukalion Verlag, Hamburg

Howard, Albert (Engl. 1948), (1979), (2005): Mein landwirtschaftliches Testament. OLV Verlag, Xanten

Howard, Louise, E. (1956): Die biologische Kettenreaktion. Boden – Kompost – Pflanzengesundheit. Sir Albert Howards Forschungen und Erfahrungen in Indien. HansGeorg Müller Verlag, Krailing

Hubendick, B. (1986): Människoekologi. Gidlunds, Stig Welinder, Department of Archaelogy, Gustavianum, Uppsala University, Sweden

Idel, Anita (2010): Die Kuh ist kein Klima-Killer – Wie die Agrarindustrie die Erde verwüstet und was wir dagegen tun können. Herausgegeben von der SchweisfurthStiftung, Munich

Jackson, Louise E. (1997): Ecology in Agriculture. Edited by Louise E. Jackson, Academic Press, San Diego. 1997. 05. Nitrogen as a Limiting Factor: Crop Acquisition of Ammonium and Nitrate, Arnold J.

Jakob, Bettina (2008): Pflanzen können mehr als angenommen http://arch.uniaktuell.unibe.ch/content/umweltnatur/2008/stickstoff/index_ger.ht ml (visited on 7/4/2016)

Janson, Arthur (1904), (1926), (2010): Auf 300 qm Gemüseland den Bedarf eines Haushalts ziehen. Anleitung zum Gemüsebau des kleinen Mannes und zur Bewirtschaftung von Schreberund Kleingärten aller Art.

Jedicke, Eckard (1989): Boden. Entstehung, Ökologie, Schutz. Ravensburger Buchverlag Otto Maier, Ravensburg

Jungk, Robert and Müllert, Norbert R. (1981), (1989): Zukunftswerkstätten. Mit Phantasie gegen Routine und Resignation. Heine Verlag, Munich

Jurzitza, Gerhard (1987): Anatomie der Samenpflanzen. Thieme Verlag, Stuttgart
Kas, Vaclav (1966): Mikroorganismen im Boden. Westarp Wissenschaftsverlag, Hohenwarsleben

Khan, S. et al. (1990): Bound ^{14}C residues in stored wheat treated with (^{14}C) Deltramethrin und their bioavailability in rats. In: Journal of Agricultural und Food Chemistry no. 38, pages 1077-1082

Kickuth, Reinhold (Editor) (1982): Die ökologische Landwirtschaft. Verlag C.F. Müller, Karlsruhe

King, Franklin Hiriam (Engl. 1911), (2004): 4000 Jahre Landbau in China, Korea und Japan. OLV Verlag, Xanten

Kollath, Werner (1977), (2001): Die Ordnung unserer Nahrung. Karl F. Haug Verlag, Heidelberg

Kollath, Elisabeth (1989): Vom Wesen des Lebendigen. Biographie Werner Kollath. Eberhard Kölle Verlag, Stuttgart

Kononowa, M. M. (1958): Die Humusstoffe des Bodens. Ergebnisse und Probleme der Humusforschung. VEB Deutscher Verlag der Wissenschaften, Berlin

Könemann, Ewald (1977): Gartenbaufibel. Braumüller Verlag, Vienna

Koepf, Herbert H. (1981): Was ist Biologisch-Dynamischer Landbau? PhilosophischAnthroposophischer Verlag, Dornach

Koepf, Herbert H. et al. (1976): Biologische Landwirtschaft. Eine Einführung in die biologisch-dynamische Wirtschaftsweise. Verlag Eugen Ulmer, Stuttgart *Koschützki, Rudolf* (Editor) (no date, ca. 1930): Rationelle Landwirtschaft in Wort und Bild. Wilhelm Andermann Verlag, Berlin und Leipzig

Krämer, Elke (2006): Leben und Werk von Prof. Dr. phil. Günther Enderlein (1872–1968). Medizinhistorische Dissertation Reichl Verlag, St. Goar. (page 202 gives information on Enderlein, Schanderl, Santo, and Rusch.)

Krasilnikov, N.A. (1958), (1961): Soil, Microorganisms and higher Plants. Washington. The National Science Foundation. Originally published by the Academy of Sciences of the USSR, Moscow

Kretschmann, Kurt and Behm, Rudolf: (1996),(revised and expanded edition 2017): Mulch total. Der Garten der Zukunft. OLV Verlag, Kevelaer

Kristensen, Petter (1996): Testikkelkreft og landbruksforurensing. Cancer Epidemiology, Biomarkers & Prevention, Jan. 1996, und Kunstgj0dsel [translator's note: I seriously doubt this word actually contains a 0, but that's how it's printed in the PDF too...] kreftfarlig, Stavanger Aftenblad 9/5, 1996

Kröplin, Bernd and Henschel, Regine C. (2016): Die Geheimnisse des Wassers. Neueste erstaunliche Ergebnisse aus der Wasserforschung. AT Verlag, Aarau

Kroymann, Jürgen (2010): Turning the Table: Plants Consume Microbes as a Source of Nutrients. In: "PLOS One. A Peer-Reviewed, Open Access Journal", 2010 (5)7: e11915, published online 201, [translator's note: I assume this is a typo for either 2001 or 2010] July 30 (visited on 6/8/2016)

Krusche, Per and Althaus, Dirk and Gabriele, Ingo (1982): Ökologisches Bauen. Bauverlag, Berlin

Lammert, F.-D. (Editor) (1989): Tiere im Boden. – In: Unterricht Biologie, volume 144, Seelze

Lau, Kurt Walter (many different years): NATÜRLICH GÄRTNERN & ANDERS LEBEN. Das Biogartenund Permakultur-Magazin (founded 1958/1988)

Lau, Kurt Walter (2017): Die Methode Siegfried Lange – oder wie man aus einem einzigen Roggenkorn mindestens zehntausend Körner ernten kann. In: NATÜRLICH GÄRTNERN & ANDERS LEBEN volume 60, number 2

Lau, Kurt Walter (1990): Die Gemüse-Getreide-Mischkultur nach Siegfried Lange. Höchsternten im rauen Mittelgebirgsklima. In: GARTEN ORGANISCH volume 33, number 5, pages 10-14

Lau, Kurt Walter (1989): Kleintierhof – ökologisch, wirtschaftlich, zukunftsweisend. Pietsch Verlag, Stuttgart

Laukötter, Gerhard (2007): Bodenbilder, Bodeninhalte, Bodenlyrik. Nümbrecht-Elsenroth-Verlag: Martina Galunder Verlag, Nümbrecht

Lewis, Thomas (1976): Das Leben überlebt – Geheimnis der Zellen. Kiepenheuer & Witsch, Cologne

Liebig, Justus von (1840), (1995): Die Chemie in ihrer Anwendung auf Agricultur und Physiologie. Reprint. Agrimedia Verlag, Holm

Liebig, Justus von (1989): Boden, Ernährung, Leben. Auswahl von Texten aus vier Jahrzehnten (found and assembled by Georg E. Siebeneicher and Wilhelm Lewicki). Pietsch Verlag, Stuttgart

Liebig, Justus von (1986): Es ist ja dies die Spitze meines Lebens. (adapted by Wolfgang von Haller) Stiftung Ökologie & Landbau, Kaiserslautern

Lovelock, James (1991): Das Gaia-Prinzip. Die Biographie unseres Planeten. Artemis, Zürich and Munich

Lutzenberger, José (2003): Das Vermächtnis. RETAP Verlag, Bonn

Margulis, Lynn (1999): Die andere Evolution. Spektrum Akademischer Verlag, Heidelberg and Berlin

Margulis, Lynn and Sagan, Dorian (1999): Leben – Vom Ursprung zur Vielfalt. Spektrum Akademischer Verlag, Heidelberg and Berlin

Margulis, Lynn und Sagan, Dorion (1993): Garden of microbial delights: a practical guide to the subvisible world. Dubuque, Iowa; Kendall/Hunt

Margulis, Lynn and Schwartz, Karlene V. (1989): Die fünf Reiche der Organismen. Spektrum der Wissenschaft Verlagsgesellschaft, Heidelberg

Markl, Julia and Hampl, Ulrich (1996): Bodenfruchtbarkeit selbst erkennen. Bodenbeurteilung mit dem Spaten. Deukalion Verlag, Holm

Meyer, Thomas (Editor) (2003): Ein Leben für den Geist: Ehrenfried Pfeiffer (1899–1961). Perseus Verlag, Basel

Meyer-Renschhausen, Elisabeth and Holl, Anne (Editor) (2000): Die Wiederkehr der Gärten. Kleinlandwirtschaft im Zeitalter der Globalisierung. Studien Verlag, Innsbruck

(Migge, Leberecht): Team of authors (1981): Leberecht Migge (1881–1935). Published by the Fachbereich Stadtund Landschaftsplanung der Gesamthochschule Kassel for the exhibition"Leberecht Migge – Gartenkultur des 20. Jahrhunderts" as part of the Bundesgartenschau Kassel, 1981

Migge, Leberecht (1932): Die wachsende Siedlung nach biologischen Gesetzen. (2nd Edition) Franckh´sche Verlagshandlung W. Keller, Stuttgart

Migge, Leberecht (1918): Jedermann Selbstversorger! Eine Lösung der Siedlungsfrage durch neuen Gartenbau. Diederichs Verlag, Jena

Minnich, Jerry (1977): The Earthworm Book. Rodale Press, Emmaus, Pennsylvania

Mollison, Bill (1989), (2016) Permakultur konkret. pala Verlag, Darmstadt

Mollison, Bill (1988), (2010): Permaculture: A Designers' Manual. Tagari Publications. German translation: Handbuch der Permakultur-Gestaltung. Permakultur-Akademie im Alpenraum

Mollison, Bill (1994): Permakultur II. Praktische Anwendungen. pala Verlag, Schaafheim

Mollison, Bill and Holmgren, David (1984): Permakultur I. Landwirtschaft und Siedlungen in Harmonie mit der Natur. pala verlag, Schaafheim

Mückenhausen, Eduard (1993): Bodenkunde. DLG-Verlag, Frankfurt/Main

Müller, Georg (1963): Bodenbiologie. VEB Gustav Fischer Verlag, Jena

Naumann, Regina (1997): Bioaktive Substanzen: die Gesundmacher in unserer Nahrung. Rowohlt Verlag, Reinbek

Nessenius, S. Eva (2009): Der Planetenembryo. Neu entdeckte Zusammenhänge zwischen Evolution und der Entstehung des Planeten Erde. Books on Demand GmbH, Norderstedt

Nestroy, Othmar (2013): Den Boden verstehen. Stocker Verlag, Graz

Nettelnburg (1995): 75 Jahre Siedlung Nettelnburg. In: Der Nettelnburger Siedler, Hamburg-Bergedorf

Nicol, Hugh (1956): Der Mensch und die Mikroben. Rohwohlt Verlag, Hamburg

NLVF-utredning (1980): Energibruk ved produksjon av matvarer i norsk jordbruk. Norges landbruksvitenskaplige forskningsrad, Oslo

Opitz, Christian (1995): Ernährung für Mensch und Erde. Grundlagen einer neuen Ethik des Essens. Hans-Nietsch-Verlag

Pauli, Fritz W. (1988): Lebendverbauung der Grenzfläche zwischen Boden und Pflanze. Leben und Stoffwechsel in der Wurzelzone höherer Pflanzen. In: GARTEN ORGANISCH volume 30, number 1, pages 14-16, number 2, pages 46-48, number 3, pages 82-84

Pauli, Fritz W. (1979): Bodenfruchtbarkeit – Gemeinschaftsleistung von Pflanzenwurzeln und Mikroorganismen. In: Ruperto Carola, 62/3. Heidelberg

Paungfoo-Lonhienne, Chanyarat et al. (2013). Rhizophagy a new dimension of plant-microbe interactions. In Frans J. de Bruijn (Ed.), Molecular Microbial Ecology of the Rhizosphere (pp. 1199-1207) Hoboken, NJ United States: Wiley-Blackwell. doi:10.1002/9781118297674.ch115

Paungfoo-Lonhienne, Chanyarat et al. (2010) Turning the table: Plants consume microbes as a source of nutrients. PLoS One, 5 7: e11915-1-e11915-11. doi:10.1371/ journal.pone.0011915

Peterson, Roger et al. (1973): Die Vögel Europas. Verlag Paul Parey, Hamburg und Berlin

Pfeifer, Sebastian (Editor), (1973): Taschenbuch für Vogelschutz. DBV Verlag, Stuttgart

Pfeiffer, Ehrenfried (1984): Chromatography applied to quality testing. Bio-Dynamic Literature, Wyoming, Rhode Island

Pfeiffer, Ehrenfried (1958): Eine qualitative chromatographische Methode zur Bestimmung biologischer Werte. In: Lebendige Erde, no date, number 5, Verlag Lebendige Erde, Darmstadt

Pfeiffer, Ehrenfried (1956): Die Fruchtbarkeit der Erde. Ihre Erhaltung und Erneuerung. Rudolf Geering-Verlag, Dornach

Pfeiffer, Ehrenfried (1942): Gesunde und kranke Landschaft. Alfred Metzner Verlag, Berlin

Pichler, Franz (2015): Raoul Francé und sein Werk zu Natur und Leben. Eine kommentierte und illustrierte Bibliographie. Trauner Verlag, Linz

Podolinsky, Alex (2008): Active Perception. Verlag Biodynamic Agriculture Australia Ltd., Bellingen NSW, Australia

Podolinsky, Alex and Hatch, Trevor (2000): Biodynamik – Landwirtschaft der Zukunft. Frumenta Verlag, CH-Arlesheim

Pollack, Gerald H. (2015): Wasser – viel mehr als H2O. VAK Verlag, Kirchzarten
Pollmer, Udo et al. (2012): Wer hat das Rind zur Sau gemacht? Rowohlt Verlag, Reinbeck

Pollmer, Udo et al. (2001): Prost Mahlzeit – Krank durch gesunde Ernährung, Kiepenheuer & Witsch, Cologne 2001

Preuschen, Gerhard (1999): Leben ist Schöpfung. Slice of Life, Erika Meyer-Borrmann & Günter Borrmann, Königslutter

Preuschen, Gerhard (1991): Ackerbaulehre nach ökologischen Gesetzen. Das Handbuch für die neue Landwirtschaft. Verlag C.F. Müller, Karlsruhe

Preuschen, Gerhard (1987): Die Kontrolle der Bodenfruchtbarkeit. Eine Anleitung zur Spatendiagnose. Stiftung Ökologie & Landbau, Kaiserslautern

Rateaver, Bargyla and Rateaver, Gylver (1994): The Organic Method Primer. The Basics. The Rateavers. San Diego, CA

Rateaver, Bargyla und Rateaver, Gylver (1993): Organic Method Primer Update. (with black-and-white photographic documentation of plant endocytosis.) The Rateavers. 9049 Covina Street, San Diego, CA

Rohde, Gustav (1957): Lehrbuch der natürlichen Kompostierung. Deutscher Bauernverlag, Berlin

Rusch, Hans Peter (1968), (2004): Bodenfruchtbarkeit. Eine Studie biologischen Denkens. OLV Verlag, Xanten 2004

Rusch, Hans Peter (1974): Mineralisation« der lebenden Substanz. In: Kultur und Politik, volume 3, no. 3, pages 12-17

Rusch, Hans Peter (1969): Über den Kreislauf der lebenden Substanz. In: Kultur und Politik, volume 24.,2/69, pages 7-16. Möschberg

Rusch, Hans Peter (1960): Über Erhaltung und Kreislauf lebendiger Substanz. In: Zeitschrift für Ganzheitsforschung, o. Jg. Heft 4, pages 50-63

Rusch, Hans Peter (1955): Naturwissenschaft von Morgen. Vorlesungen über Erhaltung und Kreislauf lebendiger Substanz. Verlag Emil Hartmann, Krailling

Rusch, Hans Peter (1951). Das Gesetz von der Erhaltung der lebendigen Substanz. In: Wiener Medizinische Wochenzeitschrift, 1951, Nr. 37 u.38

Rusch, Hans Peter (1950): Der Kreislauf der Bakterien als Lebensprinzip. In: Hippokrates, o. Jg.,Heft 21, pages 623-630

Rusch, Volker (2001): Mikrobiologische Therapie. K. F. Haug Verlag, Heidelberg
Rusch, Volker (1999): Bakterien – Freunde oder Feinde? Urania Verlag, Berlin
Samaj, Josef (Ed.) (2006): Endocytosis in Plants. Endocytosis is a crucial step in a multitude of signaling processes in plant cells. Serie Plant Cell Monographs. Englisch. Springer Verlag, Berlin

Sattler, Friedrich und Wistinghausen, von Eckard (1985): Der landwirtschaftliche Betrieb Biologisch-Dynamisch. Ulmer Verlag, Stuttgart

Sauerlandt, Walter (1958): Wir leben von der Luft! Eine historische Betrachtung über Forschungen zur biologischen Luftstickstoffbindung anlässlich der Nobelpreis-Verleihung an Prof. Dr. h.c. Artturi I. Virtanen, Institut für Biochemie, Helsinki. In: Organischer Landbau 1 Jg., Heft 1, pages 20-22

Schanderl, Hugo (1970): Bodenbakterien in neuer Sicht. Über das Entstehen von Bakterien aus pflanzlichen Zellen. In: Boden und Gesundheit. Zeitschrift für angewandte Ökologie. Nr. 68, pages 7-10

Schanderl, Hugo (1970): Wie lange noch Knöllchenbakterien? In: Boden und Gesundheit. Zeitschrift für angewandte Ökologie. Nr. 66, pages 7-10. Verlag Boden und Gesundheit, Langenburg

Schanderl, Hugo (1964): Bakterien einmal in anderer Sicht. In: Boden und Gesundheit. Zeitschrift für angewandte Ökologie. Nr. 45, pages 10-12. Verlag Boden und Gesundheit, Langenburg

Schanderl, Hugo (1950): Die Mikrobiologie des Weines. Eugen Ulmer Verlag, Stuttgart/Ludwigsburg

Schanderl, Hugo (1947): Botanische Bakteriologie und Stickstoffhaushalt der Pflanzen auf neuer Grundlage. Verlagsbuchhandlung Eugen Ulmer, Stuttgart

Schauberger, Viktor (1933), (2016): Unsere sinnlose Arbeit. Band 1 der Viktor Schauberger-Edition. J. Schauberger Verlag, Bad Ischl

Schauberger, Viktor (2006), (2012): Das Wesen des Wassers. Originaltexte, herausgegeben und kommentiert von Jörg Schauberger. AT Verlag, Aarau

Schaumann, Wolfgang (1996): Rudolf Steiner-Kurs für Landwirte. Sonderausgabe Nr. 46. Stiftung Ökologie & Landbau (SÖL), Bad Dürkheim

Scheffer, Fritz (1982): Lehrbuch der Bodenkunde – Scheffer/Schachtschabel. Ulmer Verlag, Stuttgart

Scheller, Edwin (2013): Grundzüge einer Pflanzenernährung des Ökologischen Landbaus. Verlag Lebendige Erde, Darmstadt

Scheller, Edwin (1994): Die Stickstoff-Versorgung der Pflanzen aus dem StickstoffStoffwechsel des Bodens: ein Beitrag zu einer Pflanzenernährungslehre des organischen Landbaus. Margraf Verlag, Weikersheim

Scheller, Edwin (1993): Wissenschaftliche Grundlagen zum Verständnis der Düngungspraxis im Ökologischen Landbau – Aktive Nährstoffmobilisierung und ihre Rahmenbedingungen. Ges. f. goetheanistische Forschung e.V., Dipperz.

Schinner, Franz und Sonnleitner, Renate (1996): Bodenökologie I: Mikrobiologie und Bodenenzymatik. Springer Verlag, Berlin

Schneider, Peter (2002): G. Enderleins Forschung aus heutiger Sicht. http://www.poschneider.com/enderlein/ (visited on 12/16/2016)

Schönauer, Gerhard (1979): Zurück zum Leben auf dem Lande. Wilhelm Goldmann Verlag, Munich

Schomerus, Johannes (1931): Die Bodenbedeckung – ein wertvolles Kulturverfahren. Heinrich Verlag, Dresden

Schuette, Karl H. (1965): Biologie der Spurenelemente. Ihre Rolle bei der Ernährung. Bayerischer Landwirtschaftsverlag, Munich

Schultz-Lupitz, Albert (1895), (1927): Zwischenfruchtanbau auf leichtem Boden. Berlin 1895 und Arbeiten der Deutschen Landwirtschaftsgesellschaft, volume 7; 4th edition 1927

Schuphan, Werner (1961): Zur Qualität der Nahrungspflanzen. Bayerischer Landwirtschaftsverlag, Munich

Schwarz, Max Karl (1974): Der Gärtnerhof Eine Betriebsform eigener Art im Gefüge der Landschaft. Verlag Boden und Gesundheit, Langenburg

Schwarz, Max Karl (Editor) (1947): Der Gärtnerhof – Eine Betriebsform eigener Art im Gefüge der Landschaft. With articles: "Gedanken über den künftigen Landschaftsbau im Zusammenhang mit dem Gärtnerhof (M.K. Schwarz); "Erhaltung und Steigerung der Bodenfruchtbarkeit" (W. Laatsch); "Der Gärtnerhofgedanke" (M.K. Schwarz); "Ausbildung für den Gärtnerhof" (Ernst Hagemann); "Betriebswirtschaftliche Betrachtungen über den Gärtnerhof" (Albrecht Köstlin). Verlag B. Sachse, Hamburg

Schwarz, Max Karl (1933): Ein Weg zum praktischen Siedeln. Pflugschar Verlag, Klein Vater und Sohn, Düsseldorf

Schwenk, Ernst F. (2000): Sternstunden der frühen Chemie. Verlag C. H. Beck
Seifert, Alwin (1971), (1991): Gärtnern, Ackern ohne Gift. Biederstein Verlag, Munich

Sekera, Margaret: (Sekera, Franz, 1943), (2012):Gesunder und kranker Boden. Ein praktischer Wegweiser zur Gesunderhaltung des Ackers. OLV Verlag, Kevelaer
Seymour, John (1980): Friedliches Land – Grünes Leben. Otto Maier Verlag, Ravensburg

Sheffield, Oliver Georg: (2016): Die Regenwurmfarm meines Großvaters; In: NATÜRLICH GÄRTNERN & ANDERS LEBEN: volume [translator's note: the volume isn't actually listed] no. 5, pages 52-56. Reprint from: BODEN UND GESUNDHEIT 1970, volume 67, pages 7-10

Siebeneicher, Georg E. (Editor) (1993): Handbuch für den biologischen Landbau. Das Standardwerk für alle Richtungen und Gebiete. Naturbuch Verlag, Augsburg

Smil, Vaclav (1997): Global Population und the Nitrogen Cycle. Scientific American Library, W.H. Freeman und Company, July 1997

Starka, Jiri (1968): Physiologie und Biochemie der Mikroorganismen. VEB Gustav Fischer Verlag, Jena

Statistisches Landesamt Rheinland-Pfalz (2014): Statistische Bände, Band 406, 2015 AND https://www.statistik.rlp.de/fileadmin/dokumente/nach_themen/verlag/ba- ende/band406_die_landwirtschaft_2014.pdf (visited on 12/14/2016)

Steiner, Rudolf (1979): Geisteswissenschaftliche Grundlagen zum Gedeihen der Landwirtschaft. Landwirtschaftlicher Kursus 1924 Breslau and Dornach. Rudolf Steiner Verlag, Dornach

Stellwang, Karl (1967): Kraut & Rüben. Erinnerungen und Erfahrungen eines biologischen Landwirts. Waerland Verlag, Mannheim

Thaer, Albrecht Daniel (1809-1812): Grundsätze der rationellen Landwirtschaft, 4th volume 1809-1812 Realschulbuchhandlung, Berlin

Thaer, Albrecht Daniel (1801): Einleitung der Kenntniß der englischen Landwirtschaft, 3rd volume 1798-1804, Verlag Hahn

Thomas, Lewis (1976): Das Leben überlebt. Geheimnis der Zellen. Kiepenheuer & Witsch, Cologne

Tischler, Wolfgang (1976): Einführung in die Ökologie. Gustav Fischer Verlag, Stuttgart

TLB: TBL Syntrophic Microbial Fertilizer (1995). Production in 1995 was 532 736 tons. International TBL Research Institute. Poolesville, MD 20837, China

Tompkins, Peter and Bird, Christopher (1998): Die Geheimnisse der guten Erde. John

Hamaker, 1986, quotes page 187. Omega-Verlag, Düsseldorf

Tompkins, Peter and Bird, Christopher (1989), (1992): Secrets of the Soil. pages 187-198. Arkana

Tompkins, Peter and Bird, Christopher (1977), (1990): Das geheime Leben der Pflanzen. Pflanzen als Lebewesen mit Charakter und Seele und ihre Reaktionen in den physischen und emotionalen Beziehungen zum Menschen. Fischer Verlag, Frankfurt/M.

Topp, Werner (1981): Biologie der Bodenorganismen. Verlag Quelle & Meyer, Wiebelsheim

Trolldenier, Günter (1971): Bodenbiologie. Die Bodenorganismen im Haushalt der Natur. Frankh´sche Verlagshandlung W. Keller, Stuttgart

Tvengsberg, P.M. (1992): 350 Jahre alter Finnroggen. Modellbrandrodung. N-2300 Hamar

Ungelehrt, Hans (1946): Die Selbstversorger-Siedlung – ein Weg zur wirksamen, raschen Bekämpfung des Hungers, des Wohnungsmangels und der Flüchtlingsnot.

Verband für unabhängige Gesundheitsvorsorge e.V. und Stiftung Ökologie & Landbau (2001): Vollwert-Ernährung und Öko-Landbau. Eine Einführung in die ökologische Agrarund Esskultur. Stiftung Ökologie & Landbau, Bad Dürkheim

Vershofen, Wilhelm (1946): Hauswerk und Siedlung. Albert Nauck & Co. Verlag, Berlin

Vester, Frederik (1987): Wasser = Leben. Ein kybernetisches Milieubuch mit 5 verschiedenen Wasserkreisläufen. Ravensburger Buchverlag

Virtanen, Arrturi Ilmari (1933): Über die Stickstoffernährung der Pflanzen. Anm. Sci. Fenn. A 36, Nr.12

Vitousek, Peter, M. et al. (1997): Human Alteration of the Global Nitrogen Cycle Cources und Consequences. In: Ecological Applications, 7 (3), 1997. pages 737-750. Ecological Society of Amerika

Vitusek, Peter, M. et al. (1997), (2010): World Watch. The nitrogen cycle is out of balance. Copies of the report"Human Alteration of the Global Nitrogen Cycle" can be obtained from the Ecological Society of America, 2010 Massachusetts Ave., NW, Suite 400, Washington, DC 20036 (202) 833-8773. Sept./Oct. 1997

Vogt, Gunter (2000): Entstehung und Entwicklung des ökologischen Landbaus. Verlag Stiftung Ökologie & Landbau, Bad Dürkheim

Voisin, André (1959): Boden und Pflanze – Schicksal für Mensch und Tier. BLV Verlagsgesellschaft, Munich

Voisin, André (1958): Die Produktivität der Weide. BLV Verlagsgesellschaft, Munich *Voitl, Helmut und Guggenberger, Elisabeth* (1986): Der Chroma-Boden-Test. Die Bodenqualität bestimmen, bewerten und verbessern. Ein unentbehrlicher Ratgeber für Landwirte, Berufsund Hobbygärtner. Verlag Orac, Vienna

Vollbrecht, Erich (1948): Die Stadtrand-Nebenerwerbsiedung. Elbe-Rhein-Verlag Hans Schlichting, Hamburg

Wallin, I. E. (1927): Symbionticism and the origin of species. 80. XL u. 171 pp., w. 3 pl. Baltimore Williams & Wilkins, Baltimore. Zitiert bei Schanderl, 1947, page 63

Waring, P. Alston and Teller, Walter Magnes (no date, ca. 1940): Der Erde verwurzelt. Amerikanisches Landvolk zieht Bilanz. Siebeneicher Verlag Berlin und Frankfurt/M.

Weber, Barbara; Hirn, Gerhard; Lünzer, Immo (Editor) (2000): Öko-Landbau und Gentechnik. Entwicklungen, Risiken, Handlungsbedarf. Stiftung Ökologie & Landbau, Bad Dürkheim

Weiger, Hubert und Willer, Helga (Editor) (1997): Naturschutz durch ökologischen Landbau. Stiftung Ökologie & Landbau, Bad Dürkheim

Zerluth, Josef und Gienger, Michael (2004): Gutes Wasser. Das Wesen und Wirken des Wassers. Gute Erde Verlag, Saarbrücken

Zimmermann, Werner (1975): Steine geben Brot. Verlag Ernst-Otto Cohrs, Rotenburg/Wümme

Zschocke, Anne Katharina (2016): Natürlich heilen mit Bakterien. Gesund mit Leib und Seele. AT Verlag, Aarau

Zschocke, Anne Katharina (2014): Darmbakterien als Schlüssel zur Gesundheit. Neueste Erkenntnisse aus der Mikrobiom-Forschung. Knaur Verlag, Munich
Zschocke, Anne Katharina (2012): EM Die Effektiven Mikroorganismen. Bakterien als Ursprung und Wegweiser alles Lebendigen. AT Verlag Aarau und Munich
Zschocke, Anne Katharina (2011): Die erstaunlichen Kräfte der Effektiven Mikroorganismen EM. MensSana bei Knauer, Munich

(Editor's note: All translations of foreign-language sources were done by this book's author.)

Online Sources

There have been many initiatives related to permaculture and other forms of alternative agriculture and gardening, and there are more and more each day. The "heroes of the fields" listed below are only a representative sample. It would be quite worth your while, dear readers, to use the Internet to look for similar initiatives in your area as well.

www.ackerhelden.de

„Die Essbare Stadt": www.andernach.de/de/leben_in_andernach/es_startseite.html www.bioforumschweiz.ch

www.france-harrar.de

www.gutneuenhof.de

www.humuseum.de

www.mikroveda.eu

www.permakultur-akademie.de

www.permakultur-koller.de

www.permakultur.no/ergagard (only in Norwegian)

Disclaimer:

These links are not controlled by the author or by the publisher or its employees. They are not responsible for the content of these websites. Visit the links at your own risk.

Index

Agricultural chemicals, xi, 26-28. *See also* Agrochemistry; Herbicides; Pesticides
    bound residues of, 28–30, 31
Agricultural efficiency, 18. *See also* Energy expenditure, for food production
Agricultural land, average amount per capita, 16
Agricultural practices
    "humus sapiens" model, 25, 196–204
    obsolete, 35–36
    reform of, 195–204
Agricultural production. *See* Food production
Agricultural products, new quality measures for, 202–203
Agriculture
    development of, 77, xiii
    future of, 21–22
    spiritually-based concept of, 10
Agrobiology, history of, 5–29
Agrochemistry, xviii, 1–2, 21, 25–26
    history of, 5–29

Åkerstedt, Nils, 162–166
Algae
    biomass, 59
    blue-green, 57
    as earthworm food, 157
    as human food, 141
    humus content, 135
    protein content, 141
    in rainwater tanks, 168
    soil content, 135
    use in rice cultivation, 86
Alternative agriculture, 119-129. *See also* Organic agriculture
    beginning of, 111–115
Amino acids, 26, 35
Ammonia, 10, 11
Ammonia-based fertilizers, 10, 135
Ammonia plants, 9
Animal feed, increase in use, 103
Antibiotics, 31
Antigens, 74, 82
Archaebacteria, 48, 66
Artificial fertilizer-microorganism hybrid cultures, 86
Artificial fertilizers, 24, 83

Ashes
  plant nutrients from, 2, 22
  rye cultivation in, 148–149
Atmosphere
  cytoplasm content, 93
  definition, 54–55
Augers, 170–171
Autotrophy, 14, 50, 125
Aztecs, hydroculture of, 99, 126, 139–141, 146

Bacteria
  aerobic, 79, 166, 167, 172
  anaerobic, 166, 167, 172, 173, 183
  biomass, 59
  chemoautotrophic, 4
  evolution, 15
  facultative anaerobic, 135–136
  humus content, 135
  lactic acid, 53–54, 74
  nitrogen, 74
  pathogenic, 82
  in plant decomposition, 44
  plants' absorption of, 46
  primordial, 41–42, 48
  putrefactive, 74, 135–136
  regeneration. *See* Remutation
  silt content, 138
  soil content, 135
  "stone-eating" (lithobionts), 95, 138
Bacteriology, history of, 11–12
Bacteriophages, 86
Baja California, natural humus of, 133–135
Bark compost, 181
BASF, 10
Béchamp, Pierre Jacques, 7–8, 19, 48
Beef cattle production, energy efficiency of, 100–101
Biocenosis, 93, 135
Biocides, food preservatives as, 80–81
Biodynamic agriculture, 10, 76, 128–129, 198

Biogas, 75, 76
Biological model, of agriculture, 21
Biological tillage, 61
Biomass
  bioenergy from, 76
  of edaphon, 59–60
  metabolism in humusphere, 66
Biosphere, 54–55
  economic overview of, 107–109
  humusphere component of, 66–67
  maintenance of, 65–66
  nitrogen content, 22–23
  poisoning of, 103, 190
  self-regulation of, 66–67
Blobel, Günther Klaus-Joachim, 18
Boorstein, Daniel J., xi
Bosch, Carl, 10
Bound residues, 28–30, 31
Boussingault, Jean-Baptiste, 10–11
Breast milk, toxins in, 27, 34, 81, 115
Bretscher, Mark, 40

Capra, Fritjof, 17
Carbohydrates
  absorption in plant roots, 11
  synthesis of, 74
Carbon, soil content, 52
Carbon dioxide, excessive release of, 24
Carson, Rachel, 14
Cell model, of decomposition, 46
Cellulose, 134–135
  decomposition, 74
Center for Organic-Biological Agriculture, Switzerland, 205
Chaff, 64
Chemical derivatives, 81
Chemical elements, 2
Chemical model, of agriculture and plant nutrition. *See* Agrochemistry
Chemoautotrophs, 4

Chloramphenicol, 28
Chlorella, 201
Chlorine, 93
Chlorocholine chloride (Cycocel), 30
Chlorophyll water, 71, 178–181
Chloroplasts
    in carbohydrate synthesis, 74
    DNA of, 45, 66
    in remutation, 11–12, 17, 26, 43, 46, 48
Clay, 22
Cleopatra, 138
Climate change, xi, 24, 25
Club of Rome, 16, 154
Community gardens, 154–155
Compost/composting, 35, 76–77, 81
    advantages and disadvantages, 160
    earthworm-produced, 157
    hot, 167
    large-scale plots, 155
    microorganisms in, 33
    nutritional value, 70
    process of, 160
    "surface," 171
Compost starters, 184
Crop yield. *See* Harvest yield
Cyanobacteria, 57, 58
Cycle of living material model, xvii–xviii, 46-51
    agricultural methods, 159–185
    author's personal experiences, 174–185
    chlorophyll water, 178–181
    composting, 157, 160
    kitchen-scrap porridge, 178, 185
    manure and slurry use, 166–168
    mulching, 161–166
    onion cultivation, 176–178
    organic waste reuse, 174–176
    rainwater tanks, 168
    sawdust and bark use, 181
    soilization, 170–174, 185
    wintering the edaphon, 69, 168–170
Cycle of living material theory
    comparison with mineral theory, 4, 13–14, 125–126
    definition, 36
    development of, 12
    early proponents of, 15
    first-stage nutrients, 122
    important publications about, 112
    second-stage nutrients, 122
    third-stage nutrients, 122–125, 192–193

Dahl, Jürgen, 196
Darwin, Charles, 5, 6
DDT (dichlorodiphenyltrichloroethane), 27, 28, 30, 115
Decomposition, 44
    aerobic, 75, 76–77
    anaerobic, 75, 76
    cell model of, 46
    remutation in, 44, 46
Deforestation, 102–103
Deltamethrin, 30
Diatoms, humus content, 135
Dieldrin, 29–30
Digestion, symbiotic intestinal bacteria in, 88
Dioxin, 81
Disease, cell pathology in, 128–129
Dömer, Professor, 191–192
Drugs
    bound residues of, 28
    excretion of, 31
Duckweed, 201

Earthworm husbandry, 6
Earthworms, 63
    benefits to soil, 157
    biomass, 59
    in Egyptian culture, 138
    poisoning of, 62

in symbiotic agriculture, 127–128
tunnels of, 85, 157
Ecological agriculture, xx
    comparison with technological agriculture, 122
    development of, 11
    Erga Gard Partial Garden Estate, 153–154
    limitations of, 36, 127, 192
    plant endocytosis and, 115–116
Ecological Growing (Krusche), 192
Ecology in Agriculture (Jackson), 24
Economic concept, of nature, 107–109
Economic efficiency, of agriculture, 97-110. *See also* Energy expenditure, for food cultivation
Edaphon, 31–32, 33, 55–65
    benefits to soil, 199
    biomass, 59–60
    effect of compost on, 161
    effect of food preservatives on, 81
    effect of mulching on, 161–168
    effects of technological agriculture on, 61–62
    as focus of biologically-based agriculture, 196
    functions of, 111–112
    of hydroculture systems, 62, 65
    ideal soil content of, 53, 61, 197
    living condition requirements of, 155, 157
    moisture requirements, 155, 157, 176
    nutrition for, 68, 69, 78–79, 155, 157, 168–176
    as plant nutrient source, 61–65
    topsoil formation in, 57–58
    winter maintenance, 69, 168–170
Educational programs, in organic-biological agriculture, 201–202

Egypt, ancient, 138
Elbe River Valley, 142, 143
Enderlein, Günther, 20, 112
Endocytosis, 39–43
    in animals, 39–40, 41–42
    comparison with mineral model, 41
    definition, 26, 39, 40, 50–51, 184, xii
    research about, 17
Endosymbiotic theory, 21, 42, 201
    acceptance of, 18
    comparison with mineral model, 41
    remutation and, 43–46
    research and publications about, 15–16, 112
Energy expenditure, for food production, 16–17
    among the Aztecs, 126
    food quality and, 106–107
    in gardens, 99–100, 101–102, 108
    in technological agriculture, 100–104
    in traditional agriculture, 99–102, 105
Energy sources, organic *versus* inorganic, 4
Erga, Ingvald, 153–154
Erga Gard Partial Garden Estate, Norway, 153–154
Euglena, 56
Europe, ecological agriculture in, 127
Evolution
    of agriculture, xiii
    of bacteria, 12, 15
    endosymbiosis theory of, 45
    of life on earth, xii–xiii
    of multi-cellular organisms, 66
Evolutionary biology research, 19
Exocytosis, 40, 47–48

Fenvalverate, 30
Fermentation, 74, 75
　lactic acid, 77, 165, 171
　in manure, 167
Fertilizers
　ammonia-based, 10, 135
　artificial, 24, 83, 86
　effect on food quality, 83
　effect on soil health, 32
　grass mulch as, 165
　microorganism hybrid cultures of, 86
　mineral, dominance of, 13
　nitrogen, phosphorus, potassium (NPK), 50, 84, 121
　ratio to harvest yield, 103
　soluble nitrogen release from, 24
First-stage nutrients, 122
Fish remains, as fertilizer, 155, 156
Floating gardens, 99, 137, 139, 140, 146
Food
　chemical bound residues content, 28–30
　lack of nutritional value, 107
　new quality measures for, 202–203
　pesticide content, 28–30
Food preservatives, 79–81
Food production. *See also* Energy expenditure, for food production
　land required per capita, 154
　in technological agriculture, 65
　in traditional agriculture, 114
Fossil fuels, 24
Francé, Raoul H., 9, 32–33, 55, 57, 61
Francé-Harrar, Anne, 9, 12–13, 53, 57–58, 61, 86–87, 203
　on humus moisture content, 168
Fungi, 54
　biomass, 59
　growth on food, 80
　humus content, 135
　silt content, 138

Gaia theory, 17, 20, 21, 43, 48, 58, 65–68, 112, 201
Garden-estate concept, 150–154
Gardening
　food production levels in, 99–100, 101–102, 108
　resurge in, 19
　urban, 19
Garden of Microbial Delights (Margulis and Sagan), 18, 61
"gas-eaters," 41
Genetically-modified organisms (GMOs), xix–xx, xvi–xviii, 81, 221
　horizontal gene transfer in, 34, 78, 81, 83
　proteins from, 81
Genetic revolution, 13
German agriculture, 103
Glacial water/"milk," 95, 144–145, 146
Globalization, 12–13, 19
Grain, pesticide-contaminated, 30
Grain cultivation, energy expenditure/harvest yield ratio, 105
Grass infusion, 182–183
Grass mulching method, 162–166
　large-scale applications, 165–166
Greenhouse gases, 24, 25
Green manure, 161
Green Revolution, 13, 98
Groundwater, 94
Growth hormones, 35
Gustavssen, Börje, 170–171

Haber, Fritz, 10
Hamaker, John, 17, 39–40
Hamm, W., 5, 7, 21
Harvest yield, xx
　energy requirements in, xvii
　in hydroculture, 62, 92
　losses in, 103–104
　ratio to fertilizer use, 103
　in technological agriculture, 98–99
　in traditional agriculture, 98–99, 114

Havana, Cuba, 155
Hay infusion, 182–183
Hay mulch, 170
Health, role of bacteria in, 81–83
Healthy eating, 83
Hensel, Julius, 9
Herbicides, 61–62, 103
Heterotrophism, 14, 18, 25, 50, 116, 125
Higa, Teruo, 84, 86, 171–172, 192
Hitschfeld, Oswald, 101
Hohenlohe Agricultural School, 150
Holistic models, of agriculture, xv
    biodynamic, 10, 76, 128–129, 198
Holl, Anne, 19
Holmgren, David, 16
Holzer, Sepp, 192
Horizontal gene transfer, xviii, 34, 78, 81, 83
Houseplants, chlorophyll watering of, 178–181
Howard, Albert, 11
Humans
    biomass, 59
    ecosystem, 81–83
    evolution, xii–xiii
Humus. *See also* Edaphon; Humusphere
    catastrophic decline in, 12–13
    definition, 22, 52
    early research about, 7
    effect of deforestation on, 102–103
    formation of, 74, 102
    ideal microorganism composition, 135
    layers of, 74
    mineral content, 52
    moisture content, 168
Humusphere, 51–55
    definition, 54–55, 66
    formation of, 53–55
"Humus sapiens" model, of agriculture, 25, 196–204
Hunza, hydroculture of, 143–145
Huuta agriculture, 146–150

Hydroculture, 25–26
    of the Aztecs, 99, 126, 139–141, 146
    edaphon of, 62, 65
    in the Elbe Valley, 142, 143
    fundamentals of, 95–96
    harvest yields, xvii, 62, 92
    of the Hunza, 143–145
    of the Inca, 144
    of Nile civilizations, 136–138
    rice crab agriculture, 137, 141–142
Hydrosphere, 54–55

I. G. Farben, 9
Inca, 144
Industrial agriculture, xi. *See also* Technological agriculture
Industrial wastewater, 93
Information sources, about humusphere agriculture, 159–160
    online, 251
Infusoria cultures, 135, 145–146, 168
    from grass mulch, 165
Inorganic substances, as plant nutrients, 3–4
Insects, biomass, 59
Iron, 22

Journal of Applied Ecology, 150–151

Kennedy, Margrit, 192
Kitchen-scrap porridge, 178, 185
Kononova, M. M., 52
Kraämer, Elke, 20
Krasilnikov, Nikolai Aleksandrovich, 14
Krusche, Maria, 151–153
Krusche, Per, 151–153, 192

Lactic acid bacteria, 53–54, 74
Lactic acid fermentation, 77, 165, 177
Lange, Siegfried, 62, 64, 147, 148, 149, 199
Lau, Kurt Walter, 151

Legumes
    nitrogen production in, 11
    rhizobia on, 8
Lettuce cultivation, 71, 163, 180, 182
Lime, 22
Liquid manure, 31, 62, 75
Lithobionts, 95
    silt content, 138
Lithosphere, 54–55, 66, 95
Little Humusphere Museum, Norway, 159
Livestock production, relation to plant production, 167–168
Lovelock, James, 17, 20, 192

Macromolecules, 123–125
Mad cow disease, 2
Magnesium oxide, 22
Malathion, 30
Manure
    as fertilizer, 77–79, 135
    liquid, 31, 62, 75
    microorganisms in, 33
    nitrogen, phosphorus, potassium content, 24, 75, 78
    processing of, 79, 167
    proper application of, 166–168
    storage of, 167
    toxins in, 34, 78, 183
Margulis, Lynn, 15–16, 18, 19, 20, 21, 43, 45, 48, 112, 192
Metabolites, 28
Methrin, 29–30
Methylococcus capsulatus, 201
Meyer-Renschhausen, Elisabeth, 19
Microbial cultures, 84–88
Microbial therapy, 112, 113, 115–116
Microbiology, based on the humusphere, 200–201
Microorganisms, in soil, 55, 73–89. *See also* Bacteria; Edaphon; Plankton
    beneficial, 87
    as biological quality indicator, 81–82
    cellulose-decomposing, 74
    coexistence with humans, 81–83
    effective, 4, 84–89, 184, 199, 201
    extermination of, 87
    fermentation, 74
    in hydroculture systems, 25–26
    necessary for humus health, 87–88
    pathogenic, 87–88
    plants' digestion and absorption of, 50–51
    in rainwater tanks, 168
    in soil layers, 74
    as starter preparations, 184
    symbiotic, 88
MicroVeda®, 184
Microzymas, 7–8
Milk production, 105, 151–152
Mineral fertilizers. *See also* Mineral theory
    adverse effects of, 116
Minerals
    definition, 52
    humus content, 52
    ideal soil content, 53, 61
    in soil water, 94
Mineral theory, 1–4, 198, xii, xvi
    adverse effects of, 53, 190, 196
    comparison with cycle of living material model, 4, 13–14, 125–126
    comparison with endocytosis, 41
    comparison with remutation theory, 48–49
    definition, 3–4, 8
    implications for soil health, 33–34
    need for alternatives to, 190–193
Molecular biology, 21
Mollison, Bill, 16, 100, 192
Monomorphism, 49
Mulch/mulching, 71, 77, 161–168
    Åkerstedt's grass mulch method, 162–166

with black plastic, 161
conventional methods, 161
Müller, Hans, 13, 205
Multi-cellular organisms,
    evolution of, 66
Mushrooms, chemical
    contamination of, 30
Mycorrhiza, symbiosis in, 11

Natural foods, 123
Naturopathy, xiv
Nematodes, humus content, 135
Nettelnburg, Germany, 142, 143
Nile civilizations, hydroculture in,
    136–138
Nitrates
    breakdown of, 11
    free, 24
Nitrogen
    biosphere content, 22–23
    manure content, 75
    as nutrient, 2
    rainwater content, 93
    regulation in nature, 24
    utilization in plants, 10–11
    water content, 92
Nitrogen, phosphorus, potassium
    (NPK)
    manure content of, 78
    stinging nettle liquid content, 183
    as toxins, 9
Nitrogen, phosphorus, potassium
    (NPK) fertilizers, 121
    effect on human health, 84
    rationale for use, 50
Nitrogen cycle, 24
Nitrogen fertilizers
    adverse effects of, 9, 24–25
    increase in use, 103
Nitrous oxide, 25
Nobel Prize, 18–19
Norway, garden-estate concept in,
    150–154

Norwegian Institute for Agricultural
    Research, 16–17
Nucleus, survival after cell death, 46
Nutrients
    biologically-active, 123–125
    qualitative characteristics of, 2–3

"One straw revolution," 127–128
Onion cultivation, 69, 176–178
Organelles
    DNA content, 66
    in remutation, 17, 48, 49
Organic agriculture, 119–129
    beginning of, 15–20
    as humus agriculture, 120–126
    importance of, 83
    manure use in, 166–167
Organic-biological agriculture
    educational programs in, 201–202
    foundation of, 36
Organic-biological (orbio) theory,
    definition, 198
Organic Chemistry in its Applications to
    Agriculture and Physiology
    (von Liebig), 1, 8, 33
Organic fertilizers
    chemically-contaminated, 31–32
    nitrogen content, 24
Organic materials. *See also*
    Organic waste
    absorption by plants, 23
    conversion in soil, 22–23
    ideal soil content, 53, 61
    reuse, 200
Organic Method Primer Update, The (B.
    and G. Rateaver), xvi, 40
Organic waste, 168–170
    daily production of, 182
    as edaphon nutrition, 182
    fermentation, 171–172
    kitchen-scrap porridge, 178
    nutritional value, 182
    reuse, 174–176

soilization of, 170–174, 185
    use as potting soil, 172–173
Oxygen
    production in plants, 66
    water content, 92

Paramecium caudatum, 56
Pasteur, Louis, 7, 9, 80
Pasteurization, 80
Permaculture, 150, 174
    founders of, 16, 100
Pesticides, 27, 62
    accumulation in fatty tissue, 28
    bound residues of, 28–30, 31
    DDT, 27, 28, 30, 115
    effect on food quality, 83
    excretion of, 28–29
    food content, 28–30, 83
    free residues of, 29, 30
    increase in use, 103
    Silent Spring (Carson) and, 14
Phosphoric acid, 22
Photosynthetic organisms, 4
Pirimiphos-methyl, 30
Plankton
    freshwater, 145–146
    of the humusphere. *See* Edaphon
    in hydroculture, 25–26
Plant disease, 106
    viral, 116
Plant fortifiers, 184
Plant nutrients
    absorption in plants, 36
    from ashes of burned plants, 2, 22
    edaphon as source of, 61–65
    heterotrophic nature of, 26
    water-soluble, 23
Plant nutrition. *See also* Cycle of living material; Mineral model
    conventional theories of, vxiii, 121
    new model of, 50–51, 198
    organic *versus* inorganic sources of, 3–4, 8

    research and publications about, 5–20
    water-soluble-based, 195
Plants
    oxygen production in, 66
    pesticide content, 29–30
    as primary producers, 50
    toxic substance storage in, 29
Podolinsky, Alex, 51
Polychlorinated biphenyl (PCB), 27, 28
Potash, 22
Potato cultivation
    energy expenditure/harvest yield ratio, 105
    pesticide use in, 27
    recovery from viral disease, 116
Potting soil, 163–164, 172–173
Practical agricultural research, 198–199
Pre-composting, 149
Preservatives, 79–81
Preuschen, Gerhardt, 23, 66–67
Primary producers, 41, 50
Proteins
    cellular location and movement, 18
    early theories about, 9
    edaphon content, 59
    from genetically-modified organisms, xviii, 81
    microbial production of, 201
    plants' digestion and absorption of, 11, 50–51
    plants' need for, 42–43
    plants' synthesis of, 106
Protozoa
    silt content, 138
    soil content, 135
Pseudo-hormones, 35

Radishes, pesticide content, 29–30
Rainwater, 93
Rainwater tanks, 168
Rateaver, Bargyla, 17, 40, 50–51, 112

Rateaver, G., 17
Remutation, 43–46, 47, 112
    comparison with decomposition, 67–68
    comparison with mineral theory, 48–49
    definition, 26, 44, 51
    rejection of, 45–46
    scientific confirmation of, 45
Rhizopods, humus content, 135
Rice cultivation
    algae use in, 86
    energy expenditure/harvest yield ratio, 105
    rice crab agriculture, 137, 141–142, 146
Rice production, 137
Röhrig, Carl W., 85
Root hairs
    endocytosis in, 41
    microscopic image, 110
    nutrient absorption in, 26
    relation to root surface area, 52
Roots
    in carbohydrate synthesis, 74
    endocytosis in, 41–42
    nutrient absorption in, 26, 51
    soil contact surface area of, 115
    surface area, 52
    water absorption in, 51
Rusch, Hans Peter, 13-14, 20, 23, 198. *See also* Cycle of living material model
    criticism of, 19
    microbial therapy development, 86–87, 88
    on mineralization, 113
    on nutrients, 122
    publications and lectures, 13, 15, 121, 204–205
Rusch, Kerstin, 88
Rusch, Volker, 36, 88, 112
Rye, soil contact surface area of, 52, 115

Rye cultivation
    harvest yield, 62, 64, 65
    Rye Finn's method (huuta), 146–150

Sagan, Dorion, 18
Salt ions, xx
    adverse effects on plants, 25
    dissolved, xii, 24, 26, 35, 92
    soil's lack of, 66–67
Saprophytes, 32
Sauerkraut, 71, 75, 77
Sawdust, composting of, 181
Schanderl, Hugo, 11-12, 26, 112, 192. *See also* Remutation
    development of remutation theory, 43–45, 48–49
Schneider, Peter, 19
Scientific theories and models, 3
    false, xiii–xiv, 3
    functions of, 189–190
Second-stage nutrients, 122
Seifert, Alwin, 74–75, 79
Sheffield, Oliver, 6
Silent Spring (Carson), 14
Silica, 22
Silicate, 135
Silt, 136, 138, 146
Slash-and-burn agriculture, 146–150
Slurry, 76, 78, 135
    proper application of, 166
Smil, Vaclav, 18, 24
Sodium bicarbonate, 22
Soil
    compaction, 85
    effect of conventional agriculture on, 34–35
    erosion-proof, 61
    fertility, lactic acid bacteria and, 53–54
    formation
        with lichens and mosses, 174–175
        soilization, 170–174, 185

"half-rotted," 32–33
healthy, 53, 70
humus content, 51-52. *See also* Humusphere
ideal composition of, 53, 61
ideal edaphon content, 53, 61, 203
layers of microorganisms in, 74
pesticide-contaminated, 32
pore spaces in, 61
pure, 32–33
virgin, 92, 115
Soil Fertility: A Study of Biological Thought (Rusch), 15, 204-205
Soil improvers, 184
Soilization, 170–174
  within containers, 172–173, 185
Soil water, 94
Solar energy, storage in plants, 176
Solar heating, passive, 153
Soybean cultivation, energy expenditure/harvest yield ratio, 105
Spirulina, 141, 201
Spring water, 94
Steiner, Rudolf, 10, 51, 128–129, 198
Stem cell research, 20, 66
Stinging nettle liquid, 146, 157, 168, 182–184
Stone meal, 9, 95–96
Straw, 170
Strempfer, Fritz, 150
Sulfur, 22
Surface water, 95
Symbioflor, 87
Symbionts, 53–54, 83
Symbiosis
  at cellular level, 17
  mycorrhizal, 11
  theory of, 15–16
Symbiosis control, 86–87
Symbiosis therapy, 86–87
Symbiotic agriculture, 127–128
Synthetic substances, 26–27

Technological agriculture, xx
  adverse effects on edaphon, 61–62
  comparison with ecological agriculture, 122
  development, 5
  energy expenditure in, 100–104, 105
  food production in, 65
  harvest yield in, 98–99
  history, 5–20
  specialization of, 196
Technological worldview, 191
Thaer, Albrecht Daniel, 7, 21
Third-stage nutrients, 122–125, 192–193
TLB Syntrophic Microbial Fertilizer (TLB), 86
Tomato cultivation, 100
Tompkins, Peter, 17
Topsoil. *See also* Humusphere
  formation of, 53–55
Toxins
  effect on plant metabolism, 106
  produced from anaerobic decomposition, 75–76
Traditional agriculture
  artificial water supplies in, 113
  efficiency of, 18
  energy expenditure/harvest yield ratio, 99–100
  harvest yield, 98–99
Tristan da Cunha, 155
Tvengsberg, Per Martin, 147, 149–150

University of Giessen, 191–192
University of Hanover, 165–166
Urban gardening, 154–155

Vester, Frederic, 107, 108–109
Virtanen, Artturi I., 4, 10–11, 19, 23
Vitousek, Pete, 24
Vogt, Gunther, 19, 45–46
Von Haller, Albert, 150

Von Haller, Wolfgang, 150–151, 192
Von Liebig, Justus, xxi, 1-2, 3, 4, 5, 7, 22, 196. *See also* Mineral theory
    Organic Chemistry in its Applications to Agriculture and Physiology (von Liebig), 1, 8, 33

Wallin, Ivan Emmanuel, 15–16, 45
Water, 91–96
    chlorophyll, 71, 178–181
    types of, 93–95

Watermelons, 84
Water-soluble salt ions, xii, 24–26, 35, 92
Water-soluble substances, 23, 31
Weeds, as edaphon nutrition source, 169
Wintering, of the edaphon, 69, 168–170
Winter rye, 64, 146–150, 199
Wood chips, 170
Wood lice, 110
Wood shavings, 147, 150
World hunger problem, 65

Acres U.S.A. — books are just the beginning!

Farmers and gardeners around the world are learning to grow bountiful crops profitably — without risking their own health and destroying the fertility of the soil. Acres U.S.A. can show you how. If you want to be on the cutting edge of organic and sustainable growing technologies, techniques, markets, news, analysis and trends, look to Acres U.S.A. For over 45 years, we've been the independent voice for eco-agriculture. Each monthly issue is packed with practical, hands-on information you can put to work on your farm, bringing solutions to your most pressing problems. Get the advice consultants charge thousands for . . .

- Fertility management
- Non-chemical weed & insect control
- Specialty crops & marketing
- Grazing, composting & natural veterinary care
- Soil's link to human & animal health

For a free sample copy or to subscribe, visit us online at

www.acresusa.com

or call toll-free in the U.S. and Canada

1-800-355-5313

Outside U.S. and Canada call 970-392-4464
or email info@acresusa.com